GENERAL RELATIVITY

Starting with the idea of an event and finishing with a description of the standard big-bang model of the Universe, this textbook provides a clear, concise, and up-to-date introduction to the theory of general relativity, suitable for final-year undergraduate mathematics or physics students. Throughout, the emphasis is on the geometric structure of spacetime, rather than the traditional coordinate-dependent approach. This allows the theory to be pared down and presented in its simplest and most elegant form. Topics covered include flat spacetime (special relativity), Maxwell fields, the energy–momentum tensor, spacetime curvature and gravity, Schwarzschild and Kerr spacetimes, black holes and singularities, and cosmology.

In developing the theory, all physical assumptions are clearly spelled out, and the necessary mathematics is developed along with the physics. Exercises are provided at the end of each chapter and key ideas in the text are illustrated with worked examples. Solutions and hints to selected problems are also provided at the end of the book.

This textbook will enable the student to develop a sound understanding of the theory of general relativity and all the necessary mathematical machinery.

Dr. Ludvigsen received his first Ph.D. from Newcastle University and his second from the University of Pittsburgh. His research at the University of Botswana, Lesotho, and Swaziland led to an Andrew Mellon Fellowship in Pittsburgh, where he worked with the renowned relativist Ted Newman on problems connected with *H*-space and nonlinear gravitons. Dr. Ludvigsen is currently serving as both docent and lecturer at the University of Linköping in Sweden.

GENERAL RELATIVITY

A GEOMETRIC APPROACH

Malcolm Ludvigsen

University of Linköping

CAMBRIDGE
UNIVERSITY PRESS

CAMBRIDGE
UNIVERSITY PRESS

University Printing House, Cambridge CB2 8BS, United Kingdom

One Liberty Plaza, 20th Floor, New York, NY 10006, USA

477 Williamstown Road, Port Melbourne, VIC 3207, Australia

4843/24, 2nd Floor, Ansari Road, Daryaganj, Delhi - 110002, India

79 Anson Road, #06-04/06, Singapore 079906

Cambridge University Press is part of the University of Cambridge.

It furthers the University's mission by disseminating knowledge in the pursuit of education, learning and research at the highest international levels of excellence.

www.cambridge.org
Information on this title: www.cambridge.org/9780521639767

First published 1999

A catalogue record for this publication is available from the British Library

Library of Congress Cataloging in Publication data
Ludvigsen, Malcolm. 1946–
General relativity : a geometric approach / Malcolm Ludvigsen.
p. cm.
Includes bibliographical references and index.
ISBN 0-521-63019-3 (hardbound)
1. General relativity (Physics) 2. Space and time.
3. Geometrodynamics. I. Title.
QC173.6.L83 1999
530.11 – dc21 98-37546
CIP

ISBN 978-0-521-63019-1 Hardback
ISBN 978-0-521-63976-7 Paperback

To Libby, John, and Elizabeth

Contents

Preface

A tribe living near the North Pole might well consider the direction defined by the North Star to be particularly sacred. It has the nice geometrical property of being perpendicular to the snow, it forms the axis of rotation for all the other stars on the celestial sphere, and it coincides with the direction in which snowballs fall. However, as we all know, this is just because the North Pole is a very special place. At all other points on the surface of the earth this direction is still special – it still forms the axis of the celestial sphere – but not that special. To the man in the moon it is not special at all.

Man's concept of *space* and *time*, and, more recently, *spacetime*, has gone through a similar process. We no longer consider the direction "up" to be special on a worldwide scale – though it is, of course, very special locally – and we no longer consider the earth to be at the center of the universe. We don't even consider the formation of the earth or even its eventual demise to be particularly special events on a cosmological scale. If we consider nonterrestrial objects, we no longer have the comfortable notion of being in the state of absolute rest (relative to what?), and, as we shall see, even the notion of straight-line, or rectilinear, motion ceases to make sense in the presence of strong gravitational fields.

All notions, theories, and ideas in physics have a certain domain of validity. The notion of absolute rest and the corresponding notion of absolute space are a case in point. If we restrict our attention to observations and experiments performed in terrestrial laboratories, then this is a perfectly meaningful and useful notion: a particle is in a state of absolute rest if it doesn't move with respect to the laboratory. In fact, that is implicitly assumed in much of elementary physics and much of quantum mechanics. However, it ceases to be meaningful if we move further afield. How, for example, do we define a state of absolute rest in outer space? Relative to the earth? Relative to the sun? The center of the galaxy, perhaps? Like the fanciful tribe's sacred direction, it must be given up (unless, of course, there does, in fact, exist a preferred, and physically detectable, state of absolute rest), and if we are to gain a clear, uncluttered view of the workings of nature, it should not enter in any way into our description of the laws of physics – it should be set aside along with the angels who once were needed to guide the planets round their orbits.

The setting aside of long-cherished but obsolete physical theories and notions is not always as easy as it might sound. It involves a new, less parochial, and less cozy way of looking at the world and, sometimes, new and unfamiliar mathematical structures. For example, after Galileo first cast doubt on the idea, it took physicists over 300 years to finally abandon the notion of absolute rest.

One of the purposes of this book is to describe, in as simple a way as possible, our present assumptions about the nature of space, time, and spacetime. I shall attempt to describe how these assumptions arise, their domain of validity, and how they can be expressed mathematically. I shall avoid speculative assumptions and theories (I shall not even mention string theories, and have very little to say about inflation) and concentrate on the bedrock of well-established theories.

Another purpose of this book is to attempt to express the laws of physics – at least those relating to spacetime – in as simple and uncluttered a form as possible, and in a form that does not rely on obsolete or (physically) meaningless notions. For example, if we agree that physical space contains no preferred point – an apparently valid assumption as far as the fundamental laws of physics are concerned – then, according to this point of view, physical space should be modeled on some sort of mathematical space containing no special point, for example, an affine space rather than a vector space. (A vector space contains a special point, namely the null vector.) In other words, I shall attempt to expel all – or, at least some – angels from the description of spacetime.

It is no accident that I use the word "spacetime" rather than "space and time" or "space-time." It expresses the fact that, at least as far as the fundamental laws of physics are concerned, space and time form one indivisible entity. The apparent clear-cut distinction between space and time that we make in our daily lives is simply a local prejudice, not unlike the sacred direction. If, instead of going around on bicycles with a maximum speed of 10 miles per hour, we went around in spaceships with a (relative) speed of, say, 185,999 miles per second, then the distinction between space and time would be considerably less clear-cut.

PART ONE

THE CONCEPT OF SPACETIME

1

Introduction

One of the greatest intellectual achievements of the twentieth century is surely the realization that space and time should be considered as a single whole – a four-dimensional manifold called spacetime – rather than two separate, independent entities. This resolved at one stroke the apparent incompatibility between the physical equivalence of inertial observers and the constancy of the speed of light, and brought within its wake a whole new way of looking at the physical world where time and space are no longer absolute – a fixed, god-given background for all physical processes – but are themselves physical constructs whose properties and geometry are dependent on the state of the universe. I am, of course, referring to the special and general theory of relativity.

This book is about this revolutionary idea and, in particular, the impact that it has had on our view of the universe as a whole. From the very beginning the emphasis will be on **spacetime** as a single, undifferentiated four-dimensional manifold, and its physical geometry. But what do we actually mean by spacetime and what do we mean by its physical geometry?

A point of spacetime represents an **event**: an instantaneous, pointlike occurrence, for example lightning striking a tree. This should be contrasted with the notion of a **point in space**, which essentially represents the position of a pointlike particle with respect to some frame of reference. In the spacetime picture, a particle is represented by a curve, its **world line**, which represents the sequence of events that it "occupies" during its lifetime. The life span of a person is, for example, a sequence of events, starting with birth, ending with death, and punctuated by many happy and sad events. A short meeting between two friends is, for example, represented in the spacetime picture as an intersection of their world lines. The importance of the spacetime picture is that it does not depend on the initial imposition of any notion of **absolute time** or **absolute space** – an event is something in its own right, and we don't need to represent it by (t, p), where p is its absolute position and t is its absolute time.

The geometry of spacetime is not something given *a priori*, but something to be discovered from physical observations and general physical principles. A geometrical statement about spacetime is really a statement about physics, or a relationship between two or more events that any observer would agree exists. For example, the statement that two events A and B can be connected by a light ray is geometrical in that if one observer finds it to be true then all observers will find it to be true. On the other hand, the statement that A and B occur in the same place, like lightning striking the same tree twice, is certainly nongeometrical. Certainly the tree will appear to be in the same place at each lightning strike to a person standing nearby, but not to a passing astronaut who happens to be flying by in his spaceship at 10,000 miles per hour.

Just as points and curves are the basic elements of Euclidean geometry, events and world lines are the basic elements of spacetime geometry. However, whereas the rules of Euclidean geometry are given axiomatically, the rules of spacetime geometry are statements about the physical world and may be viewed as a way of expressing certain fundamental laws of physics. Just as in Euclidean geometry where we have a special set of curves called straight lines, in spacetime geometry we have a special set of world lines corresponding to freely moving (inertial) massive particles (e.g. electrons, protons, cricket balls, etc.) and an even more special set corresponding to freely moving massless particles (e.g. photons). In the next few chapters we shall show how the geometry of spacetime can be constructed from these basic elements. Our main guide in this endeavor will be the **principle of relativity**, which, roughly speaking, states the following:

If any two inertial observers perform the same experiment covering a small region of spacetime then, all other things being equal, they will come up with the same results.

In other words, all inertial observers are equal as far as the fundamental laws of physics are concerned. For example, the results of an experiment performed by an astronaut in a freely moving, perfectly insulated spaceship would give no indication of his spacetime position in the universe, his state of motion, or his orientation. Of course, if he opened his curtains and looked out of the window, his view would be different from that of some other inertial observer in a different part of the universe. Thus, not all inertial observers are equal with respect to their *environment*, and, as we shall see, the environment of the universe as a whole selects out a very special set of inertial observers who are, in a sense, in a state of absolute rest with respect to the large-scale structure of the universe. This does not, however, contradict the principle of relativity, because this state of motion is determined by the state of the universe rather than the fundamental laws of physics.

The most important geometrical structure we shall consider is called the **spacetime metric**. This is a tensorial object that essentially determines the distance (or time) between two nearby events. It should be emphasized that the existence of a spacetime metric and its properties are derived from the principle of relativity and the behavior of light; it is therefore a physical object that encodes certain very fundamental laws of nature. The metric determines another tensorial object called the **curvature tensor**. At any given event this essentially encodes all information about the gravitational field in the neighborhood of the event. It may also, in a very real sense, be interpreted as describing the curvature of spacetime. Another tensorial object we shall consider is called the **energy–momentum tensor**. This describes the mass (energy) content of a small region of spacetime. That this tensor and the curvature tensor are related via the famous **Einstein equation**

$$G_{ab} = -8\pi T_{ab}$$

is one of the foundations of general relativity. This equation gives a relationship between the curvature of spacetime (G_{ab}) and its mass content T_{ab}. Needless to say, it has profound implications as far as the geometry of the universe is concerned.

When dealing with spacetime we are really dealing with the very bedrock of physics. All physical processes take place within a spacetime setting and, indeed, determine the very structure of spacetime itself. Unlike other branches of physics, which are more selective in their subject matter, the study of spacetime has an all-pervading character, and this leads, necessarily, to a global picture of the universe as a whole. Given that the fundamental laws of physics are the same in all regions of the universe, we are led to a global spacetime description consisting of a four-dimensional manifold, M, whose elements represent all events in the universe, together with a metric g_{ab}. The matter content of the universe determines an energy-momentum tensor, T_{ab}, and this in turn determines the curvature of spacetime via Einstein's equation.

In the same way as the curvature of the earth may be neglected as long as we stay within a sufficiently small region on the earth's surface, the curvature of spacetime (and hence gravity) may be neglected as long as we restrict attention to a sufficiently small region of spacetime. This leads to a flat-space description of nature, which is adequate for situations where gravitational effects may be neglected. The study of flat spacetime and physical processes within such a setting is called **special relativity**. This will be described from a geometric point of view in the first few chapters of this book.

To bring gravity into the picture we must include the curvature of spacetime. This leads to a very elegant and highly successful theory of gravity known as **general relativity**, which is the main topic of this book. Not only is it compatible with Newtonian theory under usual conditions (e.g. in the solar system), but it yields new effects under more extreme conditions, all of which have been experimentally verified. Perhaps the most exciting thing about general relativity is that it predicts the existence of very exotic objects known as **black holes**. A black hole is essentially an object whose gravitational field is so strong as to prevent even light from escaping. Though the observational evidence is not yet entirely conclusive, it is generally believed that such objects do indeed exist and may be quite common.

In the final part of the book we deal with cosmology, which is the study of the large-scale structure and behavior of the universe as a whole. At first sight this may seem to be a rash and presumptuous exercise with little chance of any real success, and better left to philosophers and theologians. After all, the universe as a whole is a very complicated system with apparently little order or regularity. It is true that there exist fundamental laws of nature that considerably reduce the randomness of things (e.g., restricting the orbits of planets to be ellipses rather than some arbitrary curves), but they have no bearing on the *initial conditions* of a physical system, which can be – and, in real life, are – pretty random. For example, there seems to be no reason why the planets of our solar system have their particular masses or particular distances from the sun: Newton's law of gravity would be consistent with a very different solar system. Things are, however, much less random on a very small scale. The laws of quantum mechanics, for example, determine the energy levels of a hydrogen atom independent of any initial conditions, and, on an even smaller scale, there is hope that masses of all fundamental particles will eventually be determined from some very basic law of nature. The world is thus very regular on a small scale, but as we increase the scale of things, irregularity and randomness seem to increase.

This is true up to a point, but, as we increase our length scale, regularity and order slowly begin to reappear. We certainly know more about the mechanics of the solar system than about the mechanics of human interaction, and the structure and evolution of stars is much better understood than that of bacteria, say. As we increase our length scale still further to a sufficiently large galactic level, a remarkable degree of order and regularity becomes apparent: the distribution of galaxies appears to be spatially homogeneous and isotropic. Clearly, this does not apply to *all* observers – even inertial observers. If it were true for one observer, then, because of the Doppler effect, it would not be true for another observer with a high relative speed. We shall return to this point shortly.

It thus appears that the universe is not simply a random collection of irregularly distributed matter, but is a single entity, all parts of which are in some sense in unison with all other parts. This is, at any rate, the view taken by the **standard model** of cosmology, which will be our main concern.

The universe, as we have seen, appears to be homogeneous and isotropic on a sufficiently large scale. These properties lead us to make an assumption about the model universe we shall be studying, called the **cosmological principle**. According to this principle the universe is homogeneous everywhere and isotropic about every point in it. This assumption is very important, and it is remarkable that the universe seems to obey it. The universe is thus not a random collection of galaxies, but a single unified entity. As we stated above, the cosmological principle is not true for all observers, but only for those who are, in a sense, at rest with respect with the universe as a whole. We shall refer to such observers as being **comoving**. With this in mind, the cosmological principle may be stated in a spacetime context as follows:

- Any event E can be occupied by just one comoving observer, and to this observer the universe appears isotropic. The set of all comoving world lines thus forms a congruence of curves in the spacetime of the universe as a whole, in the sense that any given event lies on just one comoving world line.
- Given an event E on some comoving world line, there exists a unique corresponding event E' on any other comoving world line such that the physical conditions at E and E' are identical. We say that E and E' lie in the same ***epoch*** and have the same universal time t.

By combining the cosmological principle in this form with Einstein's equation we obtain a mathematical model of the universe as a whole, called the standard big-bang model, which makes the following remarkable predictions:

(i) The universe cannot be static, but must either be expanding or contracting at any given epoch. This is, of course, consistent with Hubble's observations, which indicate the universe is expanding in the present epoch.

(ii) Given that the universe is now expanding, the matter density $\rho(t)$ of the universe at any universal time t in the past must have been a decreasing function of t, and furthermore there exists a *finite* number t_0 such that

$$\lim_{t \to t_0^+} \rho(t) = \infty.$$

The density of the universe thus increases as we move back in time, and can achieve an arbitrarily large value within a *finite* time. From now on we shall choose an origin for t such that $t_0 = 0$.

(iii) The $t =$ constant cross sections corresponding to different epochs are spaces of constant curvature. If their curvature is positive (a closed universe) then the universe will eventually start contracting. If, on the other hand, their curvature is negative or zero (an open universe), then the universe will continue to expand forever.

(iv) Assuming that all matter in the universe was once in thermal equilibrium, then the temperature $T(t)$ would have been a decreasing function of t and

$$\lim_{t \to 0^+} T(t) = \infty.$$

In other words, the early universe would have been a very hot place.

(v) There will have existed a time in the past when radiation ceased to be in thermal equilibrium with ordinary matter. Though not in thermal equilibrium after this time, the radiation will have retained its characteristic blackbody spectrum and should now be detectable at a much lower temperature of about 3 K. Such a cosmic background radiation was discovered by Penzias and Wilson in 1965, thus giving a very convincing confirmation of the standard model. Furthermore, this radiation was found to be extremely isotropic, thus lending support to the cosmological principle.

(vi) Using the well-tried methods of standard particle physics and statistical mechanics, the standard model predicts the present abundance of the lighter elements in the universe. This prediction has been confirmed by observation. For a popular account of this see, for example, Weinberg (1993).

A very disturbing feature of the standard model is that it predicts that the universe started with a **big bang** at a *finite* time in the past. What happened before the big bang, and what was the nature of the event corresponding to the big bang itself? Such questions are based on deeply ingrained, but false, assumptions about the nature of time. The spacetime manifold M of the universe consists, first of all, of *all* possible events that can occur in the universe. At this stage no time function is defined on M, and we do not assume that one exists *a priori*. However, using certain physical laws together with the cosmological principle, a universal time function t can be *constructed* on M. This assigns a number $t(E)$ to each event, and, by virtue of its construction, the range of t is all *positive* numbers not including zero. Thus, there simply aren't any events such that $t(E) \leq 0$, and, in particular, no event E_{BB} (the big bang itself) such that $t(E_{\mathrm{BB}}) = 0$ exists. Universal time in this sense is similar to absolute temperature as defined in statistical mechanics [see, for example, Buchdahl (1975)]. Here we start with the notion of a system in thermal equilibrium, and then, using certain physical principles, construct a temperature function T that assigns a number $T(S)$ – the temperature of S – to any system S in thermal equilibrium. The function T does not exist *a priori* but must be constructed. The range of the resulting function is all positive numbers not including zero. Thus, systems such that $T(S) \leq 0$ simply do not exist.

One of the most appealing features of the standard model is that it follows logically from Einstein's equation and the cosmological principle. Except possibly for the very early universe, we are on firm ground with Einstein's equation. However, the cosmological principle should be cause for concern. After all, the universe is not exactly isotropic and homogeneous – even on a very large scale – and deviations from isotropy and homogeneity might well imply a nonsingular universe without an initial big bang. That this cannot be the case can be seen from the **singularity theorems** of Hawking and Penrose. These theorems imply that if the universe is approximately isotropic and homogeneous in the present epoch – which is the case – then a singularity must have existed sometime in the past. A very readable account of these singularity theorems can be found in Hawking and Penrose (1996), but for the full details see Hawking and Ellis (1973).

Let us now return briefly to the principle of relativity. We have tacitly assumed that given two events on the world line of an observer (such events are said to have **timelike** separation) there is an absolute sense in which one occurred before the other. For example, I am convinced that my 21st birthday occurred before my 40th birthday. But are we really justified in assuming that "beforeness" in this sense is any more than a type of prejudice common to all human beings and therefore more a part of psychology than fundamental physics? There does, of course, tend to be a very real physical difference between most timelike-separated events. A wine glass in my hand is very different from the same wine glass lying shattered on the floor, and we would be inclined to say that these two events had a very definite and obvious temporal order. However, the physical laws governing the individual glass molecules are completely symmetric with respect to time reversal, and, though highly improbable, it would in principle be possible for the shattered glass to reconstitute itself and jump back into my hand. Of course, such events never happen in practice, at least when one is sober, but this has more to do with improbable boundary conditions than the laws of physics.

For many years it was felt that all laws of physics ought to be time-symmetric in this way. This is certainly true for particles moving under the influence of electromagnetic and gravitational interactions (e.g. glass molecules), but the discovery of weak elementary-particle interactions in the fifties has called into question this attitude. It is now known that there exist physical processes governed by weak interactions (e.g. neutral K-meson decay) that are not time-symmetric. These processes indicate that there does indeed exist a physically objective sense in which the notion of "beforeness" can be assigned to one of two timelike-separated events. Of course, temporal order can be defined with respect to an observer's environment in the universe – if $\rho(E) > \rho(E')$ then, given that the universe is expanding, we would be inclined to say that event E occurred before event E' – but the type of temporal order we are talking about here is with respect to the fundamental laws of physics. A good account of time asymmetry is given in Davies (1974).

Another surprising feature of processes governed by weak interactions is that they can exhibit a definite "handedness." However, unlike that found (on the average) in the human population, which, as far as we know, is a mere accident of evolution, the type of handedness exhibited by weakly interacting processes is universal and an integral part of the objective physical world. It can, in fact, be used to obtain a physical distinction between right-handed and left-handed frames of reference, since an experimental configuration based on a right-handed frame will, in general, yield a different set of measurements from one based on a left-handed frame. For an entertaining discussion of these ideas see Gardner (1967).

Finally, we should say something about the physical units used in this book. Clearly, nature does not care which system of units we use: the time interval between two events on a person's world line is, for example, the same whether she uses seconds or hours as the unit. We shall therefore use a system of units in terms of which the fundamental laws of physics assume their simplest form.

Let us initially agree to use a *second* as our unit of time – we'll choose a more natural unit of time later. We then choose our unit of distance to be a light-second. This is a particularly natural unit of length, since one of the fundamental laws of nature is that light always has the same speed with respect to any observer. By choosing a light-second as our unit of distance we are essentially encoding this law into our system of units. Note that, in units of seconds and light-seconds, the speed of light c is, by definition, unity.

Another feature of light, and one that forms the basis of quantum theory, is that the energy of a single photon is exactly proportional to its frequency. We use this to define our unit of energy as that of a photon with angular frequency one. In terms of this unit of energy, Planck's constant $\hbar$ is, by definition, unity.

Finally, since mass is simply another form of energy (this will be shown when we come to consider special relativity), we also measure mass in terms of angular frequency. For example, if we wish to measure the mass of a particle in units of frequency, we could bring it into contact with its antiparticle. By arranging things such that the resulting explosion consists of just two photons, the frequency of one of these photons will give the mass of our particle in units of frequency.

Since we are defining distance, energy, and mass in units of time, it is important to have a good definition of what we mean by an accurate clock. As a provisional definition, we can define a clock as simply any smoothly running, cyclical device that is unaffected by changes in its immediate environment. This, for example, rules out pendulum clocks. But how can we check that a clock is actually unaffected by changes in its environment? It is no good appealing to some other, better clock – not even the most up-to-date atomic clock, which presumably ticks away the hours in a cellar of the Greenwich observatory – as this would lead to a circular argument. There is only one certain way and that is to appeal to the properties of nature herself. Given a clock, together with the appropriate apparatus, all in a unchanging environment, it is, at least in principle, possible to determine the gravitational constant G in units of time. Recall that we are defining both mass and distance in units of time. If we now change the environment (e.g. by changing the temperature or transferring the laboratory to the moon) and the value of G remains unchanged, we can say, by definition, that we have a good clock and one unaffected by changes in its environment.

The gravitational constant is, of course, defined by $G = ar^2/m$, where a is the acceleration of a particle of mass m caused by the gravitational influence of an identical particle at a distance r. If, for example, we change our unit of time from one second to one minute, then $r \to r/60$ (one light-second $= \frac{1}{60}$ light-minutes), $m \to 60m$ (one cycle per second = 60 cycles per minute), and similarly $a \to 60a$. Thus $G \to G/(60)^2$. A particularly convenient unit of time for gravitational physics is that which makes $G = 1$. This is called a **Planck second** or a **gravitational second**. Whenever we are dealing with gravity we shall use this unit of time; otherwise, for simplicity, we shall stick with ordinary seconds.

Nature gives us many other natural time units. A particle physicist might, for example, prefer to use an electron second (a unit of time such that $m_e = 1$) or even a proton second. It is a remarkable feature of our universe that clocks reading electron, proton, and gravitational time *appear* to remain synchronous.

EXERCISES

1.1 Calculate the following quantities in terms of natural units where $c = \hbar = G = 1$: the mass and radius of the sun, the mass and spin of an electron, and the mass of a proton.

1.2 Another set of natural units is where $c = \hbar = m_e = 1$, where m_e is the mass of an electron. In terms of these units calculate the quantities mentioned in Exercise 1.1.

1.3 You have made email contact with an experimental physicist on the planet Pluto. She wishes to know your age, height, and mass, but has never heard of pounds, feet, or seconds (or any other earthly units). By instructing her to perform a series of experiments, show how this information can be conveyed.

1.4 Your friend on Pluto also wants to know your body temperature. How can this information be conveyed? (Hint: Use your knowledge of statistical mechanics, and choose units such that Boltzmann's constant is unity.)

1.5 According to the principle of relativity there is no preferred state of inertial motion. Does this conflict with the cosmological principle?

2

Events and Spacetime

2.1 Events

Lightning striking a tree, a brief encounter between friends, the battle of Hastings, a supernova explosion, a birthday party – these are all examples of events. An event is simply an occurrence at some specific time and at some specific place. Events, as we shall see, form the basic elements of the spacetime description of the universe.

The **world line** of a particle is the sequence of events that it occupies during its lifetime. Birthday parties, for example, form a particularly important set of events on any person's world line. A brief encounter between two friends is an event common to both their world lines (Fig. 2.1).

Most real events are very fuzzy affairs with no definite beginning or end. A **pointlike event**, on the other hand, is one that appears to occur instantaneously to *any* observer capable of seeing it.[†] A collision between two pointlike particle, for example, is a pointlike event. It is, of course, possible to have a nonpointlike event that appears to be instantaneous to *some* observer, but, due to the finite velocity of the propagation of light, such an event will not in general appear to be instantaneous to some other observer. We say that two pointlike events occupy the same **spacetime point** if they appear to occur simultaneously to *any* observer capable of seeing them. If this is not the case, we say that they occupy distinct spacetime points. It is, of course, possible to have two events occupying distinct spacetime points that appear simultaneously to *some* observer, but, again because of the finite velocity of the propagation of light, they will not, in general, appear simultaneous to some other observer. The set M of all spacetime points is called, not surprisingly, **spacetime**.

Since it takes four parameters, for example the coordinates (t, x, y, z) with respect to some cartesian frame of reference together with some clock, to specify the spacetime position of a event, M has the structure of a four-dimensional manifold. This statement will be made more precise later, but for now all we need to know is that, at least locally, a point of M

[†] The operational definition of a pointlike event given here is dependent on the geometric optics approximation, which is adequate for most of our purposes. Strictly speaking, pointlike events do not occur in nature.

can be specified by four parameters. For the moment, it is not important how these parameters, or coordinates, are constructed.

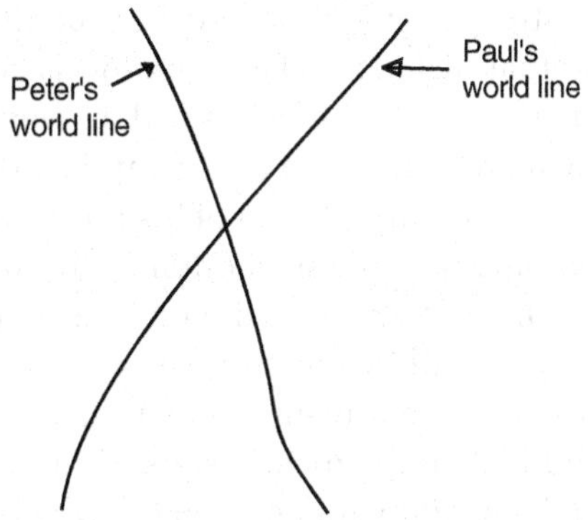

Figure 2.1. A brief encounter between Peter and Paul.

2.2 Inertial Particles

If no forces, apart from gravity, are acting on a pointlike particle, we say that it is **inertial** or in an **inertial state of motion**. We exclude gravity in this definition, because, unlike other forces like electromagnetism and friction, gravity is a property possessed by *all* material bodies, even, as we shall see, massless particles such as photons. Gravity, in other words, is a *universal* force. There exist in nature electrically neutral particles that are unaffected by electric fields, but all particles are affected in some way by gravity. Almost 400 years ago Galileo observed that inertial particles have the following remarkable property: if two such particles – one may be made of wood and the other of iron – are initially coincident with zero relative speed, then they remain coincident while in an inertial state of motion. This is not the case for noninertial particles: in the presence of air resistance a lead ball will reach the ground before a feather, if they are both dropped from the same height.

The importance of inertial particles, at least as far as the geometry of spacetime is concerned, is that their world lines form a preferred set of curves in M, which we call **inertial world lines**, or **timelike geodesics**. An inertial world line through a point (event) $p \in M$ is determined uniquely by its velocity vector (three parameters) with respect to some observer who instantaneously occupies p.

We say that an observer is in a state of inertial motion if an inertial particle that is initially at rest with respect to him remains at rest with respect to him – at least locally. By this proviso we mean that the initial acceleration of the particle with respect to the observer is zero to first order in its distance from the observer. An astronaut in a spaceship could check to see if he is in a state of inertial motion by releasing a small object directly in front of his nose; if it remains floating where it is, at least for a short while, then he can be sure that he is in a state of inertial motion. This will certainly be the case for a (nonrotating) spaceship in outer space – assuming, of course, its rocket motors are switched off. It will also be the case for a spaceship hurtling in free fall toward the surface of the earth – but only locally. This can easily be seen to be true by picturing the astronaut and the object as two freely falling particles in radial orbits toward the center of the earth. A simple calculation shows that their relative acceleration toward each other is zero to first order in d, their mutual distance. The higher-order terms are called **tidal effects** and are the telltale signs of gravity. If, however, the spaceship's rocket motors are switched on, or the spaceship is still on the launch pad, then an object that is initially in front of his nose will fall to the ground and he will have to conclude that his motion is not inertial.

This test for inertial motion is an entirely local and private affair, performed within the confines of the spaceship with the curtains drawn. Another experiment our astronaut could perform is to open the curtains and look at the distant stars: if they appear to remain fixed with respect to the window frame, then he will conclude that he is in a special state of motion with respect to the rest of the universe. It is a curious, but well-verified, fact that, in a region of spacetime where gravitational effects can be neglected, this state of motion – let us call it universal inertial motion – is equivalent to inertial motion. At first sight it may seem that this type of motion could permit acceleration along a line. However, for such a state of acceleration, even though the spaceship is not rotating, the relative positions of the distant stars would appear to change.

That universal inertial motion is not equivalent to inertial motion in the presence of gravity can be understood by imagining an observer on a large, nonrotating planet in interstellar space. By looking at the distant stars he will conclude that he is in a state of universal inertial motion, but by dropping an apple he will conclude that he is not in a state of inertial motion.

The **principle of relativity** states that if A and B are two inertial observers who perform *identical* experiments covering a region of spacetime small enough for tidal effects to be neglected, then, *all other things being equal*, they will come up with the same results. It is, of course, possible for this principle to be false, but, so far, it is supported by all experimental evidence. Accepting that the principle of relativity is true, it means that, by conducting an experiment lasting a sufficiently short interval of time and occupying a sufficiently small region of space – for example, within the confines of his spaceship – an inertial astronaut would be unable to say anything about his spacetime position, his orientation, or his state of motion. He might be floating in the depths of interstellar space or, equally well, hurtling toward the surface of the earth and about to crash. In this sense, all spacetime points are equal, all inertial world lines through any given spacetime point are equal, and all directions through any given spacetime point are equal.

If, however, our astronaut conducts a more extended experiment, occupying a larger region of space and lasting a longer time, such that tidal effects become detectable, then he will be able to say something about the gravitational field in his immediate spacetime neighborhood. The results of an experiment conducted in a spaceship in interstellar space will now be different from those of an otherwise identical experiment conducted in a spaceship about to crash into the surface of the earth.

2.3 Light and Null Cones

A sudden explosion at a point on the surface of the sun (e.g. a solar flare) is experienced by an observer on earth – let us call him Peter – as a sudden flash; that is, the photons from the explosion that reach Peter's eyes all arrive at the same time. The same applies, of course, to any other type of sudden, localized event, for example, a supernova explosion. If this were not the case then the visible universe would look very different from what it does. If the photons arrived at different times, then what in reality is a sudden event would appear to an observer on earth as a long drawn-out affair. If, for example, blue, high-energy photons arrived before red, low-energy photons, then Peter would see two performances, the first in blue and the next in red. This is, in fact, the case for the electrons produced by the explosion: high-energy (fast) electrons arrive before low-energy (slow) electrons, in the same way as a jet plane arrives at its destination before a ship. But photons, no matter what their energy is, all arrive at the same time. In other words, a pointlike event is always seen as a pointlike event.

Photons are thus very special particles with properties very different from massive particles like electrons and cricket balls. To see just how special they are, consider the following properties possessed by photons, and contrast them with properties possessed by, for example, cricket balls:

- **(i)** Some of the photons seen by Peter will have been emitted by atoms moving toward him and some by atoms moving in the opposite direction – and yet they all arrive at the same instant.
- **(ii)** Photons of different energy all arrive at the same instant.
- **(iii)** Some of the photons may have been absorbed and instantaneously reemitted by atoms during their journey to Peter, and yet they still arrive at the same instant as those photons which have had a clear run. Their instant of arrival is thus independent of the motion of the atoms that interrupt their journey. This is, of course, assuming the atoms lie in a direct line between Peter and the sun.
- **(iv)** Any observer unlucky enough to instantaneously occupy the same spacetime point as the explosion (i.e. whose world line passes through this point) would consider himself to be at the center of the explosion's wave front in the sense that all photons produced by the explosion would appear to diverge away from him – no matter what his state of motion happens to be.

Fortunately, this apparently unreasonable behavior of photons, which is totally at odds with the notions of absolute time and absolute space, can be understood in terms of a spacetime picture by assuming that photon world lines – we shall call these null rays or null curves – have the following properties:

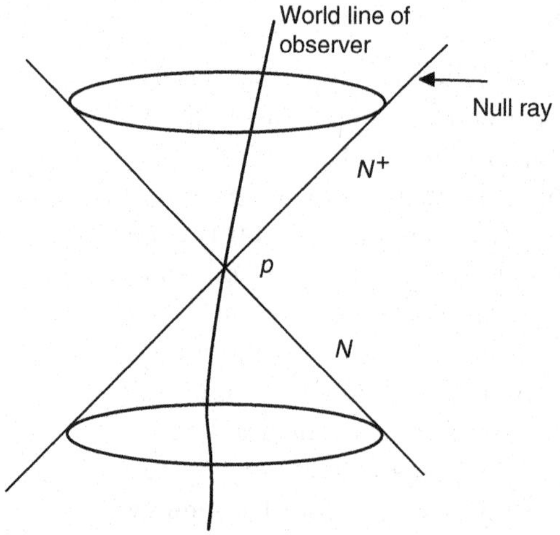

Figure 2.2. A null cone $N(p)$ with vertex p is independent of the state of motion of an observer whose world line passes through p.

- The set of all null rays through any given spacetime point $p \in M$ generate a unique three-dimensional cone $N(p)$ in M with p as its vertex. It is important to note that $N(p)$ is not defined with respect to any particular observer: given the point p, $N(p)$ is an absolute, observer-independent, geometrical structure in M. It has a future component $N^+(p)$ and a past component $N^-(p)$ (Fig. 2.2).
- The world line of any observer (not necessarily inertial) who instantaneously occupies p (i.e. whose world line passes through p) lies inside $N(p)$, and each null ray of $N(p)$ defines a unique point on his celestial sphere. This means that there are a sphere's worth of null rays through p and, given any two observers who instantaneously occupy p, there exists a one-to-one correspondence between points on their celestial spheres where two points correspond if they define the same null ray. By the principle of relativity there are no preferred null rays through p, and hence, at least locally, the set of null rays through p will be spherically symmetric with respect to *any* observer instantaneously occupying p.

Figure 2.3. A shower of photons emitted at point O by particles with world lines l and l' will generate a positive null cone through P. Peter's world line will intersect this null cone in a single point P.

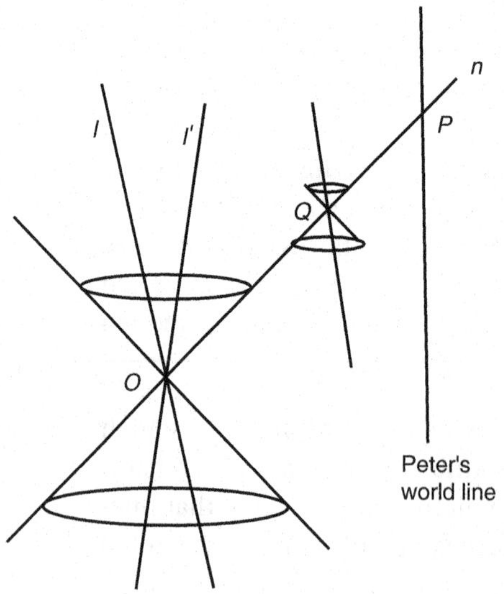

In terms of **null cones**, an observation of an explosion on the sun can be interpreted as follows: As Fig. 2.3 illustrates, the explosion at (spacetime) point O produces a shower of photons, all of which lie in the future null cone, $N^+(O)$, of O. These photons will have been emitted by atoms in various states of motion (e.g. atoms with world lines l and l' that pass through O), but they will all be contained in $N^+(O)$.

Let l_P be Peter's world line. Since l_P is one-dimensional and $N^+(O)$ is three-dimensional and both lie in a four-dimensional space M, they will intersect in a single point P. Points O and P

will lie on one null ray, n, which lies in $N^+(O)$. At P, Peter will see a sudden flash in direction defined by n, the direction of the sun from his vantage point. If any atom whose world line intersects n in a point Q absorbs a photon from the explosion and then immediately emits a shower of photons, they will form null rays in $N^+(Q)$, but only those with direction n will reach Peter and will be seen by him at point P.

In what follows we shall assume that all observers carry a clock, a light source that can emit photons of various frequencies, and a photon detector that is able to determine the frequency of a detected photon – all made to some standard specification. We shall not assume that the clocks are correlated in any way – this would almost be tantamount to introducing absolute time – nor shall we assume that our observers carry rulers or yardsticks. Such pieces of apparatus are not particularly useful for measuring distances on a astronomical scale.

Using his clock, an observer can assign a number, $t(p)$, called the **proper time**, to each point p on his world line, where $t(p)$ is the reading on his clock when he occupies p. Note that each individual observer has his own proper time and, as yet, we have defined no relationship between the proper times of different observers.

EXERCISES

2.1 Write down an expression for the relative acceleration between two inertial particles as they fall radially toward the center of the earth.

2.2 You are sitting in a freely falling spaceship. Describe an experiment that will indicate the strength of the gravitational field in your immediate vicinity.

2.3 Let S be the set of events at which two distinct events, p and q, are *seen* to occur simultaneously (i.e., to an observer instantaneously occupying $x \in S$, p and q appear to occur at the same instant). If S is nonempty, what is its dimension? If S is empty and p appears to occur before q to some observer, show that p appears to occur before q for all observers.

PART TWO

FLAT SPACETIME AND SPECIAL RELATIVITY

3

Flat Spacetime

Let us now restrict attention to a region of spacetime where gravitational tidal effects may be neglected. Such a region may extend for many light-years in interstellar space or the confines of a freely falling spaceship over an interval of a few seconds near the surface of the earth. In this chapter we shall take this region to be effectively infinite, so it is perhaps better to imagine it lying in the depths of interstellar space, well away from any gravitational influences. We shall also restrict attention to inertial particles and inertial observers and represent their world lines by straight lines. The reason for this will soon be apparent.

3.1 Distance, Time, and Angle

Our intrepid observers, Peter, Paul, and their new friend Pauline, now find themselves in the pitch blackness of interstellar space, and in order to amuse themselves – and also to discover the secrets of spacetime – they communicate by means of light rays or, equivalently, photons. Let us say that Paul emits a photon, which is received by Peter. In general, the photon's frequency according to Paul will be different from that according to Peter. This is, of course, just the Doppler effect in operation. If, however, the transmitted and received frequencies are the same whenever the experiment is performed, then Peter will say that his friend Paul has zero relative speed. By repeating this procedure but in the reverse order, Paul will say that Peter has zero relative speed – if this were not the case, then the principle of relativity would be contradicted. If their relative speeds are zero in this sense, we say that their world lines are **parallel**. An equivalent way of defining parallel world lines is this: if Paul sends out two flashes of light at times t' and $t' + \Delta t'$ (according to his clock), which are received by Peter at times t and $t + \Delta t$ (according to his clock), then their world lines are parallel if $\Delta t' = \Delta t$ (see Fig. 3.1). This means that if n pulses are sent by Paul in time $\Delta t'$ (transmitted frequency is $n/\Delta t'$), then n pulses are received by Peter in the same time $\Delta t = \Delta t'$ (received frequency is $n/\Delta t =$ transmitted frequency).

If Pauline's world line is parallel to Paul's, then the principle of relativity implies that her world line is also parallel to Peter's. We thus have an equivalence relation, namely parallelism, between inertial world lines.

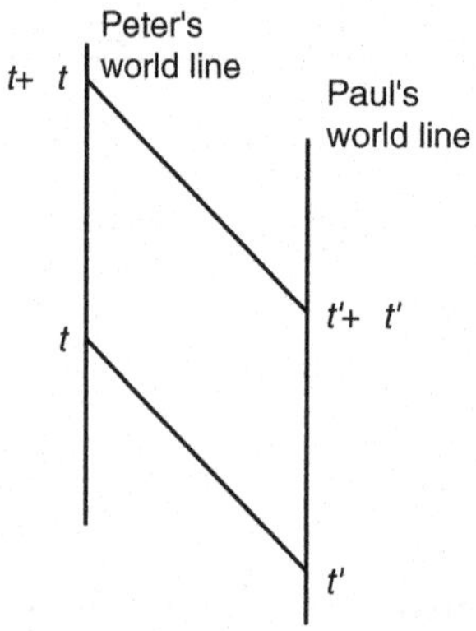

Figure 3.1. Paul sends out light signals at times t' and $t' + \Delta t'$, which are received by Peter at times t and $t + \Delta t$.

Let us now *define* what we mean by the "distance" between two parallel observers. If Peter emits a photon at time t_1 (according to his clock), which is reflected by Paul and then received by Peter at a later time t_2 (according to his clock) (see Fig. 3.2), then Peter will say that Paul is $d = (t_2 - t_1)/2$ light-seconds away – assuming, of course, that they adopt a second as their unit of time.

By the principle of relativity, this experiment will give the same result whenever it is performed, and also if performed in the reverse order, that is, Peter's distance from Paul will be the same as Paul's distance from Peter.

The same experiment enables Paul to synchronize his clock with Peter's: when Peter's photon arrives, Paul simply resets his clock to read $t_1 + d = (t_2 + t_1)/2$. Paul will know the value of d from a previous experiment, and the value of t_1 can be encoded in the photon (or, equivalently, light beam) sent by Peter. Using the principle of relativity, we see that once their clocks are synchronized in this way they will remain synchronous (this can be checked by repeating the experiment) and that Peter's clock will be synchronous with Paul's. Furthermore, if Pauline synchronizes her clock with Paul's, then her clock will be synchronous with Peter's. What we have constructed here is a universal time function applicable to any equivalence class of parallel world lines: the universal time of a point p is the proper time, $t(p)$, on the parallel world line that passes through p, that is, $t(p)$ is the reading on a parallel observer's clock when he (or she) instantaneously occupies p. We say that any two points p and q are **synchronous** if $t(p) = t(q)$. Given an inertial world line, we have an equivalence class of parallel world lines, and given a point p on the world line, we have an equivalence class of points that are synchronous with p. This equivalence class of points forms a three-dimensional plane through p (Fig. 3.3).

Figure 3.2. Peter sends out a light signal at time t_1, which is reflected by Paul and received by Peter at a later time t_2.

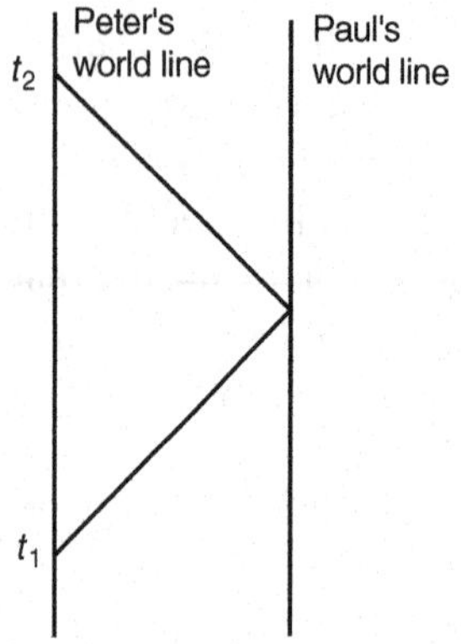

If Peter emits a photon at time t_1, which bounces off Paul at point q – *or any other observer who happens to occupy* q – and receives it back again at a later time t_2, he will say that the point (event) q occurred at time $(t_2 + t_1)/2$ and at distance $(t_2 - t_1)/2$ (Fig. 3.4).

What will Peter see if he looks out of the window of his spaceship? He will see two stationary points of light, one of which is Paul and the other Pauline. If he finds that the angle between these two points of light is θ, we define the angle between displacements OP and OQ (see diagram) to be θ (Fig. 3.5).

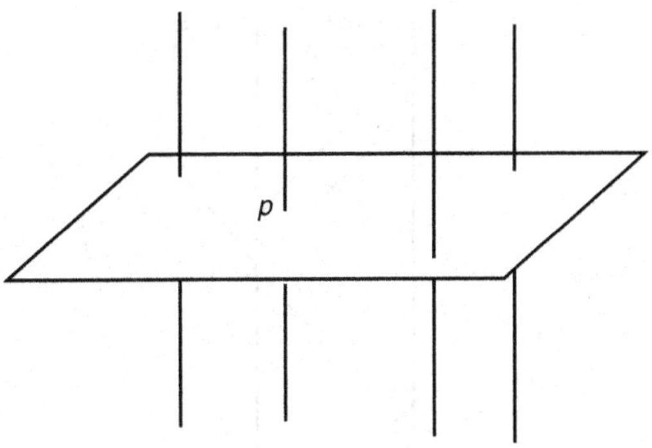

Figure 3.3. The set of all events synchronous with p forms a three-plane.

3.2 Speed and the Doppler Effect

If a rogue observer – let us call her Pat – flies past Peter and then, $\Delta t'$ seconds later (according to her clock), sends him a photon, he will receive it $\Delta t = K\Delta t'$ seconds later (according to his clock), where $K \geq 1$ (why?). If, a while later, she passes Paul and repeats the experiment with him, the results will be exactly the same: Paul will see Pat flying past him and then, $K\Delta t'$ seconds later, he will receive her photon (Fig. 3.6).

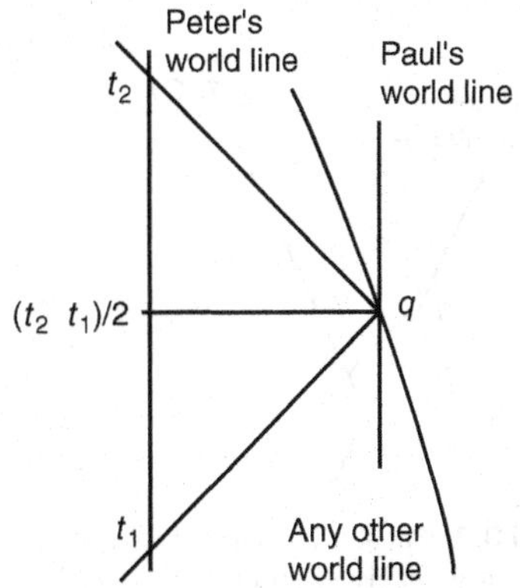

Figure 3.4. Event q occurs at time $(t_2 + t_1)/2$ and at distance $(t_2 - t_1)/2$, according to Peter.

Figure 3.5. If the angle between Paul and Pauline as seen by Peter is θ, we define the angle between the spacetime displacements OP and OQ to be θ.

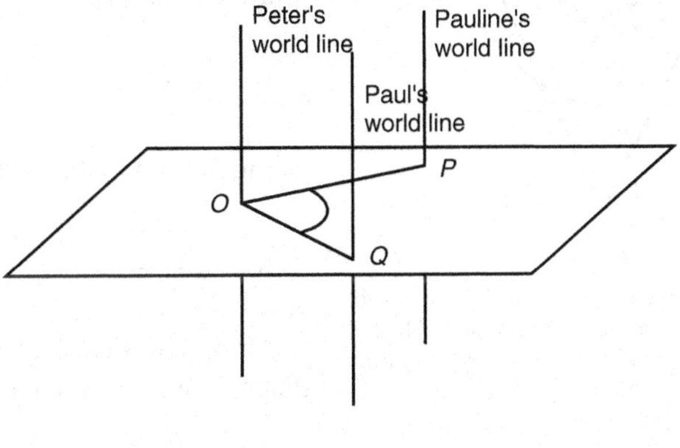

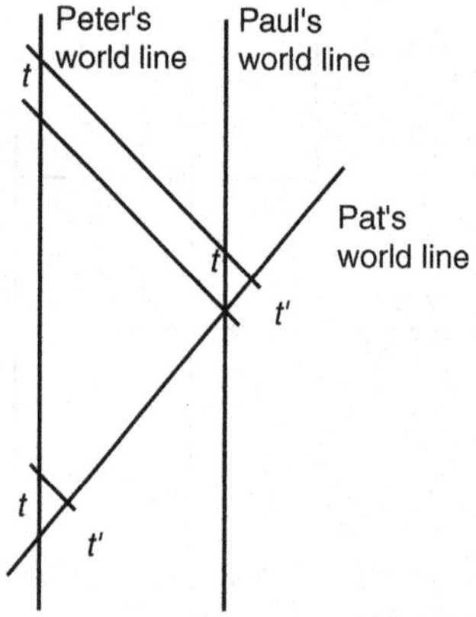

Figure 3.6. At time $\Delta t'$ after passing Peter, Pat sends out a light signal. She then repeats the same procedure after passing Paul.

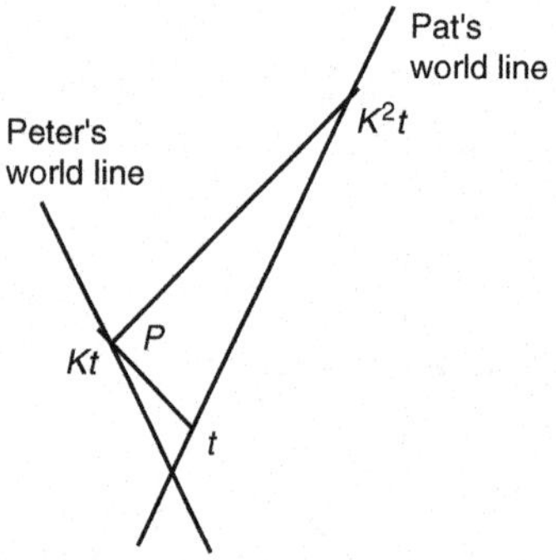

Figure 3.7. Pat sends out a light signal at time t, which is received and reflected by Peter at time Kt and then received by Pat at time K^2t.

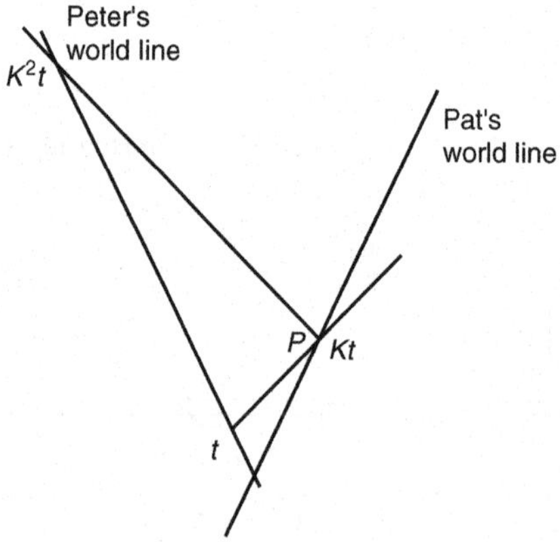

Figure 3.8. Peter sends out a light signal at time t, which is received and reflected by Pat at time Kt and then received by Peter at time K^2t.

Since Peter and Paul are parallel (zero relative speed), Paul's time interval $K\Delta t'$ will not change when passed on to Peter. Thus, if at *any* point on her world line after meeting Peter, Pat sends Peter a photon and then another one $\Delta t'$ seconds later, Peter will receive the second one $K\Delta t'$ seconds after receiving the first. Furthermore, by the principle of relativity, the reverse will also be true: If at *any* point on his world line after meeting Pat, Peter sends Pat a photon and then another one $\Delta t'$ seconds later, Pat will receive the second one $K\Delta t'$ seconds after receiving the first. The number K, which is called the **Bondi factor** (Bondi 1980) between Peter and Pat, thus depends only on their *relative* motion. Let us now find the relation between their Bondi factor K and their relative speed away from each other.

Consider Fig. 3.7, where all times refer to their respective world lines and Peter and Pat set their clocks to zero when they meet. Pat will say that point P on Peter's world line occurred at time $(K^2+1)\Delta t/2$ (according to her universal time) and at distance $(K^2 - 1)\Delta t/2$, and hence that Peter's relative speed is

$$v = \frac{K^2 - 1}{K^2 + 1}. \tag{3.1}$$

Using the fact that $K \geq 1$ this gives

$$K = \sqrt{\frac{1+v}{1-v}}. \tag{3.2}$$

Similarly, Fig. 3.8 shows that Peter will say that Pat's relative speed is also given by equation (3.1), which, of course, it should be by the principle of relativity.

Since $K \geq 1$, equation (3.1) implies $0 \leq v < 1$. This means that the relative speed (as defined above) between two inertial observers must always be less than the speed of light, which, as we are measuring distance in units of lightseconds, is equal to one. At first sight it may seem that this result flies in the face of common sense. After all, what if we had three observers, A, B, and C (Fig. 3.9) with relative speeds $v_{AB} = v_{BC} = \frac{3}{4}$? Surely v_{AC}, the relative speed between A and C, should be $\frac{3}{2}$.

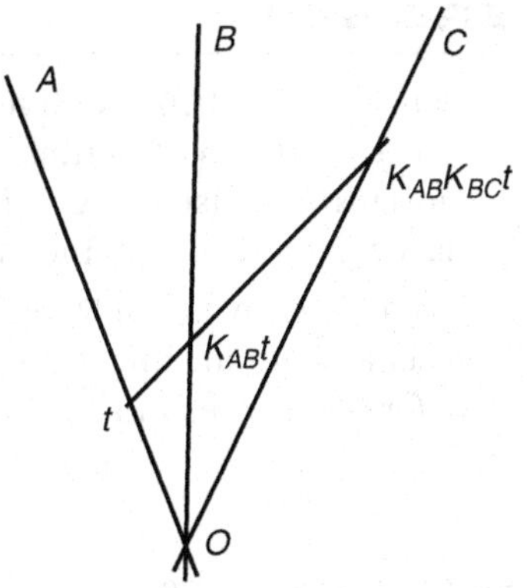

Figure 3.9. Observer A sends out a light signal at time t, which is received by B at time $K_{AB}t$ and then by C at time $K_{AB}K_{BC}t$.

However, common sense in this instance is based on an ingrained notion of absolute time and space. That v_{AC} is not $\frac{3}{2}$ but $\frac{24}{25}$ can be seen as follows: Let the Bondi factors be K_{AB}, K_{BC}, and K_{AC}. Then

$$K_{AC} = K_{AB}K_{BC} = \sqrt{\left(\frac{1+v_{AB}}{1-v_{AB}}\right)\left(\frac{1+v_{BC}}{1-v_{BC}}\right)}.$$

Hence

$$v_{AC} = \frac{K_{AC}^2 - 1}{K_{AC}^2 + 1} = \frac{v_{AB} + v_{BC}}{1 + v_{AB}v_{BC}}, \tag{3.3}$$

and thus if $v_{AB} = v_{BC} = \frac{3}{4}$ we get $v_{AC} = \frac{24}{25}$.

Let us now derive the Doppler shift between Peter and Pat, assuming that they have relative speed v and relative Bondi factor

$$K = \sqrt{\frac{1+v}{1-v}}.$$

If Peter sends Pat n pulses of light in Δt seconds (transmitted frequency $f_{\text{tr}} = n/\Delta t$), Pat will receive n pulses of light in $K\Delta t$ seconds [received frequency $f_{\text{re}} = n/(K\Delta t)$]. Hence

$$f_{\text{re}} = K^{-1} f_{\text{tr}} = \sqrt{\frac{1-v}{1+v}}\, f_{\text{tr}}. \tag{3.4}$$

EXERCISES

3.1 Consider Fig. 3.10. Observer O will say that event p occurred at distance $x = (t_2 - t_1)/2$ at time $t = (t_2 + t_1)/2$, and observer O' will say that p occurred at distance $x' = (t'_2 - t'_1)/2$ at time $t' = (t'_2 + t'_1)/2$. If O and O' have relative speed v, find the transformation between (t, x) and (t', x').

3.2 We have derived the Doppler shift between two observers *after* they meet and are moving away from each other. Derive the Doppler shift *before* they meet and are moving toward each other.

3.3 Pat says goodbye to Peter, gets into her spaceship, and travels off at half the speed of light. After one year she becomes homesick and returns to her friend Peter, again at half the speed of light, returning when she is two years older. How much older will Peter be on her return?

3.4 A train, 100 yards long, hurtles through a tunnel, 50 yards long. Show that if its speed is $\sqrt{3}/2$, then the driver (at the front) will leave the tunnel just as the guard (at the back) is entering it, according to a stationary observer in the tunnel.

3.5 If the angle between two stars is θ according to an observer occupying event p, will it also be θ according to any other observer occupying p?

Figure 3.10. O and O' are two inertial observers who set their clocks to zero when they meet. All t-values refer to proper time.

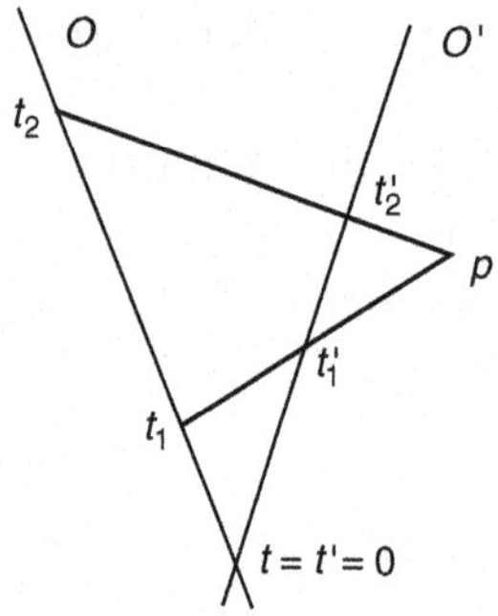

4

The Geometry of Flat Spacetime

In this chapter we shall consider the intrinsic geometric structure of flat spacetime M. In particular, we shall use the results of the previous chapter to show that M has a natural *physically defined* affine structure (i.e., the notion of a displacement vector makes sense) and, most importantly, that the space of displacement vectors possesses a natural, physically defined metric.

4.1 Spacetime Vectors

A spacetime displacement is simply an ordered pair of spacetime points. We write $OP = -PO$ and $OP + PQ = OQ$. If P lies inside $N(O)$, the null cone of O, we say that OP is **timelike**, in which case O and P lie on the world line of an observer or a massive particle. There are two types of timelike displacements: if P lies in the future of O (according to an observer whose world line passes through O and P), we say that OP is **future-pointing**; otherwise, of course, we say it is past-pointing. If P lies on $N(O)$, we say that OP is **null**, in which case O and P lie on a null ray or the world line of a massless particle. Again, there are two types of null vectors, future-pointing and past-pointing, depending on whether they lie on $N^+(O)$ or $N^-(O)$. If OP is neither timelike nor null, that is, if P lies outside $N(O)$, we say that OP is **spacelike** (Fig. 4.1).

For a timelike displacement OP, the points O and P lie on a world line l with proper time t. If Q also lies on l and $t(Q) - t(O) = \alpha[t(P) - t(O)]$, we write $OQ = \alpha\, OP$. If OP and $O'P'$ lie on parallel world lines l and l' and $t(P) - t(O) = t'(P') - t'(O')$, we say that OP and $O'P'$ are **equivalent** (Fig. 4.2).

So far, this notion of equivalence applies only to timelike displacements, but, as Fig. 4.3 shows, it has a natural extension to other types of displacements.

We say that PQ and $P'Q'$ are equivalent if the timelike displacements OP and OQ are equivalent to $O'P'$ and $O'Q'$. Physically, O is a possible spacetime meeting point of two observers whose world lines pass through P and Q, respectively.

All displacements can thus be partitioned into equivalence classes where, for example, the equivalence class $\mathbf{v} = \{OP\}$ represents the class

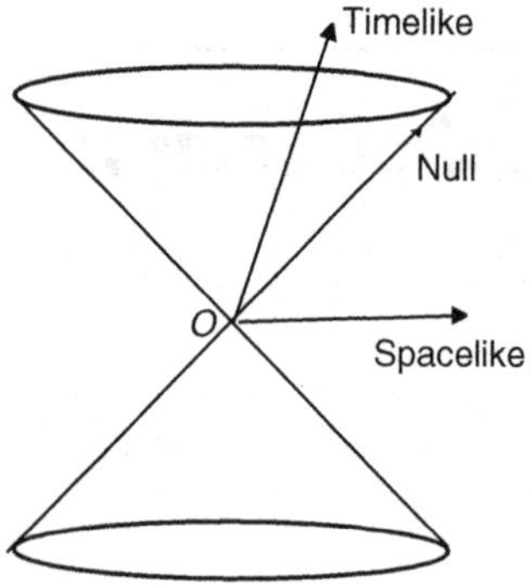

Figure 4.1. Timelike, null and spacelike displacements.

of all displacements equivalent to OP. We call such equivalence classes **spacetime vectors** and define addition and scalar multiplication by $\alpha \boldsymbol{v} = \{\alpha\, OP\}$ and $\boldsymbol{v} + \boldsymbol{w} = \{OP + PQ\}$, where $\boldsymbol{v} = \{OP\}$ and $\boldsymbol{w} = \{PQ\}$. The set of all spacetime vectors forms a four-dimensional vector space, which we call V. A given point O selects a unique representative OP from the equivalence class (vector) $\boldsymbol{x} = \{OP\}$. Thus, *given some origin point* O, we can identify V with the space of displacements that start or end at O, and simply write $\boldsymbol{x} = OP$. We say that $\boldsymbol{x}$ is the *position* vector of P with respect to origin point O. We shall sometimes indicate this by writing

$$P = O + \boldsymbol{x}. \tag{4.1}$$

Mathematically speaking, what we have shown is that flat spacetime M is an **affine space** modeled on the vector space V. A formal mathematical definition of an affine space can be found in almost any book on linear algebra, for example Halmos (1974), but a particularly readable definition is contained in Crampin and Pirani (1986).

4.2 The Spacetime Metric

If OP is timelike and $t(P) - t(O) = 1$, we say that $\boldsymbol{v} = \{OP\}$ is a **four-velocity vector**. In other words, a four-velocity vector connects two points O and P that are one second (proper time) apart and are such that P lies to the future of O. If an observer's world line passes through O and P, we say that he has four-velocity $\boldsymbol{v}$. If he sets his clock to zero at O, it will read time α at a point Q with position vector $\alpha\boldsymbol{v}$ with respect to origin point O.

Figure 4.2. Displacements OP and $O'P'$, lying on parallel world lines, are equivalent if the time interval between O and P is equal to that between O' and P'.

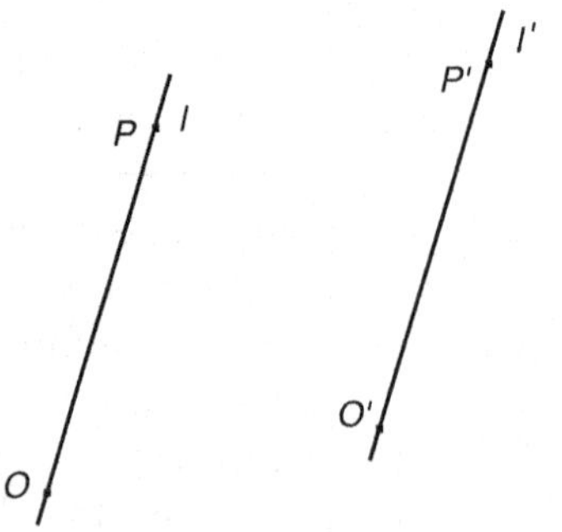

Let us now take some origin point O and some four-velocity vector $\boldsymbol{v} = OP$. The set of all vectors $\boldsymbol{x} = OQ$ where Q is synchronous with O (according to an observer with four-velocity $\boldsymbol{v}$) forms a three-dimensional subspace $\boldsymbol{v}^{\perp}$

of V. If the angle between $\boldsymbol{e} = OQ$ and $\boldsymbol{e}' = OQ'$ is 90°, we say that $\boldsymbol{e}$ and $\boldsymbol{e}'$ are **orthogonal**. Similarly, if the distance between O and Q is equal to one (i.e. one light-second), we say that $\boldsymbol{e}$ is a **unit vector**. (See the previous chapter for the *physical* definition of the distance between two synchronous points and the angle between two synchronous displacements.) The set of all unit vectors describes a closed two-surface S in $\boldsymbol{v}^\perp$ corresponding to all synchronous points one light-second distant from O. Notice that if $\boldsymbol{e}$ is a unit vector, then the vectors $\boldsymbol{v} + \boldsymbol{e}$ and $\boldsymbol{v} - \boldsymbol{e}$ are null and correspond to antipodal directions according to an observer with four-velocity $\boldsymbol{v}$.

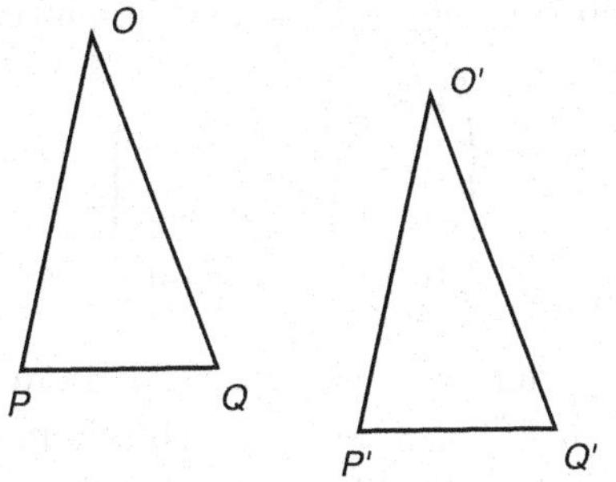

Figure 4.3. If OP and OQ are equivalent to $O'P'$ and $O'Q'$, we say that PQ is equivalent to $P'Q'$.

If $(\boldsymbol{e}_1, \boldsymbol{e}_2, \boldsymbol{e}_3)$ is a right-handed orthogonal basis of unit vectors in $\boldsymbol{v}^\perp$, we say that $(\boldsymbol{e}_0, \boldsymbol{e}_1, \boldsymbol{e}_2, \boldsymbol{e}_3)$, where $\boldsymbol{v} = \boldsymbol{e}_0$, is a spacetime **ON basis**. [It is important to note that this basis is not orthonormal (ON) with respect to some scalar product – such a scalar product exists, but we still have to show it – but is orthonormal in a *physical* sense.]

Throughout this book we will employ the convention that Greek indices take values 0 to 3 and use the Latin letters i, j, k, l, m, and n to indicate indices that take values 1 to 3. Thus, for example, we write $(\boldsymbol{e}_\alpha)$ for $(\boldsymbol{e}_0, \boldsymbol{e}_1, \boldsymbol{e}_2, \boldsymbol{e}_3)$ and $(\boldsymbol{e}_i)$ for $(\boldsymbol{e}_1, \boldsymbol{e}_2, \boldsymbol{e}_3)$. The components w^α of a vector $\boldsymbol{w}$ with respect to $(\boldsymbol{e}_\alpha)$ are given by

$$\boldsymbol{w} = w^0\boldsymbol{e}_0 + w^1\boldsymbol{e}_2 + w^2\boldsymbol{e}_2 + w^3\boldsymbol{e}_3 \equiv w^\alpha \boldsymbol{e}_\alpha.$$

Here and throughout the rest of the book we employ the **Einstein summation convention** whereby we automatically sum over repeated indices, one upstairs and the other downstairs. For example, using this convention, we see that the projection of $\boldsymbol{w}$ in $\boldsymbol{v}^\perp$ can be written as $\boldsymbol{W} = w^i\boldsymbol{e}_i$. In general, we shall use capital letters to indicate the vectors in $\boldsymbol{v}^\perp$. If a point P has position vector $\boldsymbol{x}$, we define its coordinates with respect to $(\boldsymbol{e}_\alpha)$ and the origin point O to be (x^α). We shall sometimes write $(x^\alpha) = (t, x, y, z)$.

Let us now define a natural quadratic form h on $\boldsymbol{v}^\perp$ by $h(\boldsymbol{e}_i, \boldsymbol{e}_j) = 1$ if $i = j$ and $h(\boldsymbol{e}_i, \boldsymbol{e}_j) = 0$ if $i \neq j$. This does not *necessarily* imply that $h(\boldsymbol{e}, \boldsymbol{e}) = 1$ for any (physical) unit vector $\boldsymbol{e}$ or that $h(\boldsymbol{e}, \boldsymbol{e}') = 0$ for any two (physically) orthogonal vectors $\boldsymbol{e}$ and $\boldsymbol{e}'$. However, it can easily be seen that if these conditions did not hold, then the principle of relativity would be contradicted. Using h, we now define an indefinite quadratic form g on V by $g(\boldsymbol{e}_0, \boldsymbol{e}_0) = 1$ [i.e. $g(\boldsymbol{v}, \boldsymbol{v}) = 1$], $g(\boldsymbol{e}_0, \boldsymbol{e}_i) = 0$, and $g(\boldsymbol{e}_i, \boldsymbol{e}_j) = -h(\boldsymbol{e}_i, \boldsymbol{e}_j)$. Note that $g(\boldsymbol{e}_\alpha, \boldsymbol{e}_\beta) = \eta_{\alpha\beta}$, where $\eta_{\alpha\beta} = 0$ for $\alpha \neq \beta$, $\eta_{00} = 1$, and $\eta_{11} = \eta_{22} = \eta_{33} = -1$. In matrix form this gives

$$\eta_{\alpha\beta} = \begin{pmatrix} 1 & 0 & 0 & 0 \\ 0 & -1 & 0 & 0 \\ 0 & 0 & -1 & 0 \\ 0 & 0 & 0 & -1 \end{pmatrix}. \tag{4.2}$$

In terms of g, a vector $\mathbf{x}$ is null iff $g(\mathbf{x}, \mathbf{x}) = 0$, or, in component form, iff $\eta_{\alpha\beta}x^\alpha x^\beta = 0$. This is important, since the notion of nullness is independent of any arbitrary choice of basis. We have thus demonstrated that there *exists* a quadratic form, namely g, such that $\mathbf{x}$ is null iff $g(\mathbf{x}, \mathbf{x}) = 0$.

Let us now consider the question of the uniqueness of a quadratic form g subject to this geometric, basis-independent condition. Let g' be any other quadratic form satisfying this condition. In component form this means

$$\eta_{\alpha\beta}x^\alpha x^\beta = 0 \quad \text{iff} \quad \eta'_{\alpha\beta}x^\alpha x^\beta = 0,$$

which implies $\eta'_{\alpha\beta} = \alpha\eta_{\alpha\beta}$ for some number α. Our quadratic form is thus defined up to a factor α. Let us take one such quadratic form, g, and two four-velocity vectors $\boldsymbol{v}$ and $\boldsymbol{v}'$. If $g(\boldsymbol{v}, \boldsymbol{v}) \neq g(\boldsymbol{v}', \boldsymbol{v}')$, then the principle of relativity would be contradicted. We must therefore have $g(\boldsymbol{v}, \boldsymbol{v}) = g(\boldsymbol{v}', \boldsymbol{v}')$ for any two four-velocity vectors, and we can thus fix the arbitrary factor α by demanding $g(\boldsymbol{v}, \boldsymbol{v}) = 1$ for *any* four-velocity vector $\boldsymbol{v}$. This gives a unique quadratic form g, identical to the one constructed above using an ON basis. g is called the **spacetime metric**, and, as we shall see, it forms the most important structural element in spacetime physics. Expressed in the form of a theorem, we have:

Theorem 4.1 *There exists a unique quadratic form g, called the spacetime metric, such that*

- **(i)** *$g(\mathbf{x}, \mathbf{x})$ equals 1, 0 or -1 according as $\mathbf{x}$ is a four-velocity vector, a null vector, or a unit spacelike vector.*
- **(ii)** *If $\mathbf{v}$ is a four-velocity vector and $\mathbf{e} \in \mathbf{v}^\perp$, then $g(\mathbf{v}, \mathbf{e}) = 0$.*
- **(iii)** *If $\mathbf{e}_1, \mathbf{e}_2 \in \mathbf{v}^\perp$ are orthogonal, then $g(\mathbf{e}_1, \mathbf{e}_2) = 0$.*

To ease the notation a little we shall sometimes write $\boldsymbol{w} \cdot \boldsymbol{u}$ instead of $g(\boldsymbol{w}, \boldsymbol{u})$. Note that if $\boldsymbol{w}$ and $\boldsymbol{u}$ lie in $\boldsymbol{v}^\perp$, then $\boldsymbol{w} \cdot \boldsymbol{u}$ is equal to *minus* the usual dot product on the Euclidean space $\boldsymbol{v}^\perp$. In terms of the metric, a vector $\mathbf{x}$ is timelike, null, or spacelike according as $\mathbf{x} \cdot \mathbf{x} > 0$, $\mathbf{x} \cdot \mathbf{x} = 0$, or $\mathbf{x} \cdot \mathbf{x} < 0$. A vector $\boldsymbol{v}$ is a four-velocity vector if it is future-pointing and $\boldsymbol{v} \cdot \boldsymbol{v} = 1$.

The metric g is an example of a **bilinear mapping**. Since such mappings occur frequently in spacetime physics, it is perhaps wise to say something about them now. A mapping Q, linear in each of its slots, which takes an ordered pair of vectors, $\boldsymbol{a}$ and $\boldsymbol{b}$, and gives a number $Q(\boldsymbol{a}, \boldsymbol{b})$ is said to be bilinear. If Q is symmetric in that $Q(\boldsymbol{a}, \boldsymbol{b}) = Q(\boldsymbol{b}, \boldsymbol{a})$, it is called a **quadratic form**; if it is antisymmetric in that $Q(\boldsymbol{a}, \boldsymbol{b}) = -Q(\boldsymbol{b}, \boldsymbol{a})$, it is called a **2-form.** If there exist vectors $\boldsymbol{A}$ and $\boldsymbol{B}$ such that

$$Q(\boldsymbol{a}, \boldsymbol{b}) = (\boldsymbol{A} \cdot \boldsymbol{a})(\boldsymbol{B} \cdot \boldsymbol{b}),$$

Q is said to be **simple.** It can easily be seen that any bilinear form can be expressed as a linear combination of simple bilinear forms:

$$Q(\boldsymbol{a}, \boldsymbol{b}) = \sum_i c_i(\boldsymbol{A}_i \cdot \boldsymbol{a})(\boldsymbol{B}_i \cdot \boldsymbol{b}).$$

The **trace** of Q is defined to be

$$\operatorname{tr} Q = \sum_i c_i \boldsymbol{A}_i \cdot \boldsymbol{B}_i.$$

For example, given an ON-basis $(\boldsymbol{e}_\alpha)$, the metric can be written as

$$\begin{aligned} g(\boldsymbol{a}, \boldsymbol{b}) = {} & (\boldsymbol{e}_0 \cdot \boldsymbol{a})(\boldsymbol{e}_0 \cdot \boldsymbol{b}) - (\boldsymbol{e}_1 \cdot \boldsymbol{a})(\boldsymbol{e}_1 \cdot \boldsymbol{b}) \\ & - (\boldsymbol{e}_2 \cdot \boldsymbol{a})(\boldsymbol{e}_2 \cdot \boldsymbol{b}) - (\boldsymbol{e}_3 \cdot \boldsymbol{a})(\boldsymbol{e}_3 \cdot \boldsymbol{b}) \end{aligned}$$

[this can checked by showing that $g(\boldsymbol{e}_\alpha, \boldsymbol{e}_\beta) = \eta_{\alpha\beta}$] and thus

$$\operatorname{tr} g = (\boldsymbol{e}_0 \cdot \boldsymbol{e}_0) - (\boldsymbol{e}_1 \cdot \boldsymbol{e}_1) - (\boldsymbol{e}_2 \cdot \boldsymbol{e}_2) - (\boldsymbol{e}_3 \cdot \boldsymbol{e}_3) = 4.$$

In terms of any basis $(\boldsymbol{e}_\alpha)$ the components of Q are given by $Q_{\alpha\beta} = Q(\boldsymbol{e}_\alpha, \boldsymbol{e}_\beta)$. If Q is a quadratic form, then $Q_{\alpha\beta} = Q_{\beta\alpha}$; if it is a 2-form, then $Q_{\alpha\beta} = -Q_{\beta\alpha}$. A bilinear mapping Q defines (and is defined by) a linear mapping $\hat{Q} : V \rightarrow V$ such that

$$Q(\boldsymbol{a}, \boldsymbol{b}) = \hat{Q}(\boldsymbol{a}) \cdot \boldsymbol{b}.$$

The components of $\hat{Q}$ are given by $\hat{Q}(\boldsymbol{e}_\alpha) = Q_\alpha{}^\beta \boldsymbol{e}_\beta$ and hence

$$Q_{\alpha\gamma} = Q_\alpha{}^\beta \boldsymbol{e}_\beta \cdot \boldsymbol{e}_\gamma = Q_\alpha{}^\beta \eta_{\beta\gamma}.$$

Note that we do not put a hat on $Q_\alpha{}^\beta$, since the positions of its indices indicate that it is a linear mapping rather than a bilinear form. For completeness we also define $Q^{\alpha\beta}$ by

$$Q^{\sigma\beta} \eta_{\sigma\alpha} = Q_\alpha{}^\beta.$$

It can easily be checked that

$$Q(\boldsymbol{a}, \boldsymbol{b}) = Q^{\alpha\beta} (\boldsymbol{e}_\alpha \cdot \boldsymbol{a})(\boldsymbol{e}_\beta \cdot \boldsymbol{b})$$

and hence

$$\operatorname{tr} Q = Q^{\alpha\beta} \boldsymbol{e}_\alpha \cdot \boldsymbol{e}_\beta = Q^{\alpha\beta} \eta_{\alpha\beta} = Q_\beta{}^\beta. \tag{4.3}$$

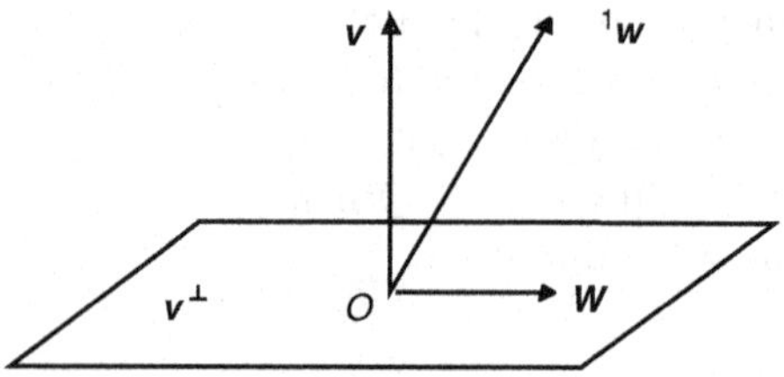

Figure 4.4. Given a four-velocity vector $\boldsymbol{v}$, any other four-velocity vector $\boldsymbol{w}$ can be written as $\gamma^{-1}\boldsymbol{w} = \boldsymbol{v} + \boldsymbol{W}$ where $\boldsymbol{W} \in \boldsymbol{v}^{\perp}$ and γ is some positive number.

Let us now consider the physical meaning of $\boldsymbol{v} \cdot \boldsymbol{w}$ where $\boldsymbol{v}$ and $\boldsymbol{w}$ are the four-velocities of two observers O_v and O_w whose world lines meet at some point O. By writing

$$\gamma^{-1}\boldsymbol{w} = \boldsymbol{v} + \boldsymbol{W} \tag{4.4}$$

(see Fig. 4.4) where $\boldsymbol{W} \in \boldsymbol{v}^{\perp}$, we see that O_v will say that O_w has traveled a distance W in unit time (i.e. the length of his four-velocity vector $\boldsymbol{v}$). Here W is the length of $\boldsymbol{W}$ and is given by $W^2 = -\boldsymbol{W} \cdot \boldsymbol{W}$. W is thus O_w's speed and $\boldsymbol{W}$ is his velocity according to O_v. Since $\boldsymbol{w} \cdot \boldsymbol{w} = 1$, equation (4.2) gives

$$\gamma = \frac{1}{\sqrt{1 - W^2}}. \tag{4.5}$$

Also, by taking the dot product of (4.4) with $\boldsymbol{v}$ we find that

$$\boldsymbol{v} \cdot \boldsymbol{w} = \gamma = \frac{1}{\sqrt{1 - W^2}}. \tag{4.6}$$

Since this equation is one of the cornerstones of special relativity, we express it in the form of a theorem:

Theorem 4.2 *If two observers have four-velocities* $\boldsymbol{v}$ *and* $\boldsymbol{w}$, *then their relative speed* W *satisfies*

$$\boldsymbol{v} \cdot \boldsymbol{w} = \frac{1}{\sqrt{1 - W^2}}.$$

We thus have a relationship between $\boldsymbol{v} \cdot \boldsymbol{w}$ and the relative speed of O_v and O_w. In particular this shows that

$$\boldsymbol{v} \cdot \boldsymbol{w} \geq 1 \tag{4.7}$$

with equality only when $\boldsymbol{v} = \boldsymbol{w}$.

If we wished we could now simply use this relationship to find the connection between $\boldsymbol{v} \cdot \boldsymbol{w}$ and the Bondi factor K between O_v and O_w, but let us do this in a more interesting and instructive way.

Consider Fig. 4.5 where $\boldsymbol{p}$ is a null vector. By the definition of the Bondi factor, we have

$$K\boldsymbol{w} = \alpha \boldsymbol{p} + \boldsymbol{v} \tag{4.8}$$

which, since $\boldsymbol{p}$ is null, gives $K^2 - 2\gamma K + 1 = 0$ and hence

$$\gamma = \frac{K^2 + 1}{2K}. \tag{4.9}$$

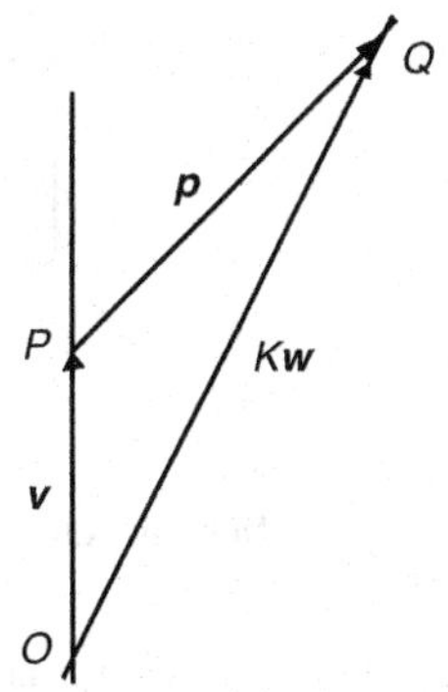

Figure 4.5. Since $\boldsymbol{v}$ is a four-velocity vector, points O and P are one unit of time apart. Thus points O and Q are K units of time apart, and since $\boldsymbol{w}$ is a four-velocity vector, $OQ = K\boldsymbol{w}$.

Let us now return to our discussion of the Doppler effect. In the previous chapter we showed that the frequency of a given photon is not absolute but depends on the motion of the observer, or, in other words, his four-velocity. This raises the question whether there exists an absolute, geometric object associated with a given photon that determines its frequency with respect to any observer. If a photon's null ray intersects O_v's world line at O, there will exist a unique future-pointing null vector $\boldsymbol{p}$ that points along the null ray such that

$$\omega(\boldsymbol{v}) = \boldsymbol{p}\cdot\boldsymbol{v}, \qquad (4.10)$$

where $\omega(\boldsymbol{v})$ is the photon's (angular) frequency with respect to O_v. An observer $O_{v'}$ with four-velocity $\boldsymbol{v}'$ such that $\boldsymbol{p}\cdot\boldsymbol{v} = \boldsymbol{p}\cdot\boldsymbol{v}'$ will have the same configuration with respect to the photon as O_v, and hence $\omega(\boldsymbol{v}') = \omega(\boldsymbol{v}) = \boldsymbol{p}\cdot\boldsymbol{v}'$. For any other observer O_w there will exist a Doppler shift given by

$$\omega(\boldsymbol{w}) = K^{-1}\omega(\boldsymbol{v}). \qquad (4.11)$$

However, using equation (4.8) we see that

$$\boldsymbol{p}\cdot\boldsymbol{w} = K^{-1}\boldsymbol{p}\cdot(\alpha\boldsymbol{p} + \boldsymbol{v}) = K^{-1}\boldsymbol{p}\cdot\boldsymbol{v} = K^{-1}\omega(\boldsymbol{v})$$

and hence equation (4.10) holds for all four-velocities. We thus have the following important theorem:

Theorem 4.3 *Every photon has an associated future-pointing null vector $\boldsymbol{p}$, tangent to its world line, such that $\omega(\boldsymbol{v}) = \boldsymbol{p}\cdot\boldsymbol{v}$, where $\omega(\boldsymbol{v})$ is its frequency with respect to an observer with four-velocity $\boldsymbol{v}$.*

The vector $\boldsymbol{p}$ is called the photon's **four-momentum.**

According to Planck's law, the energy of a photon is given by $E(\boldsymbol{v}) = \omega(\boldsymbol{v})$ (recall that we are using units such that $\hbar = 1$) and hence

$$E(\boldsymbol{v}) = \boldsymbol{p}\cdot\boldsymbol{v}. \qquad (4.12)$$

Physically, the mapping $\boldsymbol{v} \rightarrow E(\boldsymbol{v})$ is defined only on the set of four-velocity vectors [$E(\boldsymbol{v})$ is the energy of the photon according to an observer with four-velocity $\boldsymbol{v}$], but, as equation (4.12) shows, it has a natural linear extension to the whole of V and may thus be considered as a *linear* mapping that takes any vector $\boldsymbol{u} \in V$ and gives a number $E(\boldsymbol{u}) = \boldsymbol{p}\cdot\boldsymbol{u}$. By

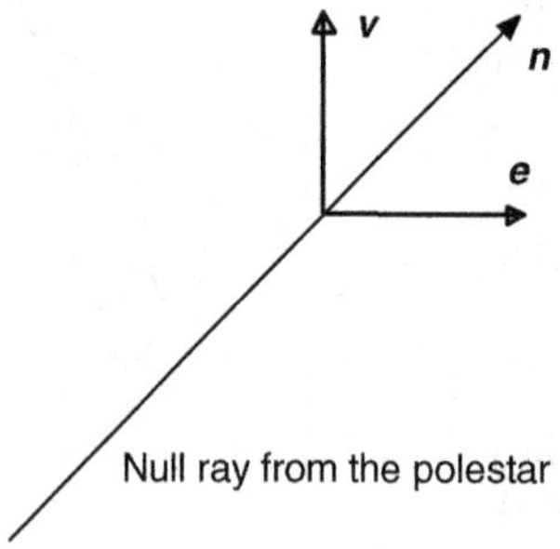

Figure 4.6. The vector $\boldsymbol{n}$ is tangent to a null ray from the polestar.

linearity we mean, of course, that

$$E(\alpha \boldsymbol{u}_1 + \beta \boldsymbol{u}_2) = \alpha E(\boldsymbol{u}_1) + \beta E(\boldsymbol{u}_2).$$

Example 4.1 Suppose the angular distance between the polestar and, let us say, Vega is θ according to Earthy, sitting on the earth. Let us calculate the corresponding angle, $\hat{\theta}$, according to Speedy, who speeds past with a speed V in the direction of the polestar.

Let $\boldsymbol{n}$ and $\boldsymbol{n}'$ be tangent to null rays from the polestar and Vega, and let $\boldsymbol{v}$ and $\hat{\boldsymbol{v}}$ be the four-velocities of Earthy and Speedy. Choosing $\boldsymbol{n}$ and $\boldsymbol{n}'$ such that $\boldsymbol{n} = \boldsymbol{v} + \boldsymbol{e}$ and $\boldsymbol{n}' = \boldsymbol{v} + \boldsymbol{e}'$ where $\boldsymbol{e}$ and $\boldsymbol{e}'$ are unit spacelike vectors orthogonal to $\boldsymbol{v}$ (see Fig. 4.6), we have

$$\boldsymbol{n} \cdot \boldsymbol{n}' = 1 + \boldsymbol{e} \cdot \boldsymbol{e}' = 1 - \cos\theta. \tag{4.13}$$

Similarly, writing $\alpha\boldsymbol{n} = \hat{\boldsymbol{v}} + \hat{\boldsymbol{e}}$ and $\alpha'\boldsymbol{n}' = \hat{\boldsymbol{v}} + \hat{\boldsymbol{e}}'$ where $\hat{\boldsymbol{e}}$ and $\hat{\boldsymbol{e}}'$ are unit, spacelike vectors orthogonal to $\hat{\boldsymbol{v}}$, we have

$$\alpha\boldsymbol{n} \cdot \hat{\boldsymbol{v}} = \alpha'\boldsymbol{n}' \cdot \hat{\boldsymbol{v}}' = 1 \tag{4.14}$$

and

$$\alpha\alpha'\boldsymbol{n} \cdot \boldsymbol{n}' = 1 + \hat{\boldsymbol{e}} \cdot \hat{\boldsymbol{e}}' = 1 - \cos\hat{\theta}. \tag{4.15}$$

Equations 4.13 and 4.15 give

$$\alpha\alpha'(1 - \cos\theta) = 1 - \cos\hat{\theta}. \tag{4.16}$$

Since Speedy is traveling *toward* the polestar (i.e. in the opposite direction to $\boldsymbol{e}$), we have

$$\hat{\boldsymbol{v}} = \gamma(\boldsymbol{v} - V\boldsymbol{e}), \tag{4.17}$$

where $\gamma^{-2} = 1 - V^2$. Since $\boldsymbol{n} = \boldsymbol{v} + \boldsymbol{e}$ and $\boldsymbol{n}' = \boldsymbol{v} + \boldsymbol{e}'$, equation (4.17) gives

$$\boldsymbol{n} \cdot \hat{\boldsymbol{v}} = \gamma(1 + V) \quad \text{and} \quad \boldsymbol{n}' \cdot \hat{\boldsymbol{v}} = \gamma(1 + V\cos\theta). \tag{4.18}$$

Putting this into equation (4.14), we get

$$\alpha\alpha' = \frac{1 - V}{1 + V\cos\theta}, \tag{4.19}$$

and finally, substituting into equation (4.16), we get the required result:

$$\cos\hat{\theta} = \frac{\cos\theta + V}{1 + V\cos\theta}. \tag{4.20}$$

Note that $\hat{\theta} < \theta$ for $V > 0$, that is, according to Speedy, Vega is closer to the polestar. Indeed, if $v \to 1$ then $\hat{\theta} \to 0$.

Example 4.2 Referring to Example (4.1), if the frequency of light from Vega is ω according to Earthy, what will it be according to Speedy?

Since $\boldsymbol{p}$ is proportional to $\boldsymbol{n}'$, $\boldsymbol{n}'\cdot\boldsymbol{v} = 1$ and $\boldsymbol{p}\cdot\boldsymbol{v} = \omega$, we have $\boldsymbol{p} = \omega\boldsymbol{n}'$. The frequency according to Speedy is thus given by $\hat{\omega} = \omega\boldsymbol{n}'\cdot\hat{\boldsymbol{v}}$. Equation (4.18) now gives

$$\begin{aligned}\hat{\omega} &= \omega\gamma(1 + V\cos\theta)\\ &= \omega\frac{1 + V\cos\theta}{\sqrt{1 - V^2}}.\end{aligned} \tag{4.21}$$

Note that this reduces to the Doppler red-shift equation for $\theta = \pi$.

4.3 Volume and Particle Density

Consider a mapping ε that takes an ordered quadruple of four-vectors $(\boldsymbol{u}_0, \boldsymbol{u}_1, \boldsymbol{u}_2, \boldsymbol{u}_3)$ and gives a number $\varepsilon(\boldsymbol{u}_0, \boldsymbol{u}_1, \boldsymbol{u}_2, \boldsymbol{u}_3)$ such that:

(i) ε is linear in each of its slots, that is,

$$\varepsilon(\alpha\boldsymbol{u}_0 + \beta\boldsymbol{u}_0', \boldsymbol{u}_1, \boldsymbol{u}_2, \boldsymbol{u}_3) = \alpha\varepsilon(\boldsymbol{u}_0, \boldsymbol{u}_1, \boldsymbol{u}_2, \boldsymbol{u}_3) + \beta\varepsilon(\boldsymbol{u}_0', \boldsymbol{u}_1, \boldsymbol{u}_2, \boldsymbol{u}_3)$$

with similar relations for the other three slots.

(ii) Under an odd permutation of $(\boldsymbol{u}_0, \boldsymbol{u}_1, \boldsymbol{u}_2, \boldsymbol{u}_3)$, $\varepsilon(\boldsymbol{u}_0, \boldsymbol{u}_1, \boldsymbol{u}_2, \boldsymbol{u}_3)$ changes sign; for example,

$$\varepsilon(\boldsymbol{u}_0, \boldsymbol{u}_1, \boldsymbol{u}_2, \boldsymbol{u}_3) = -\varepsilon(\boldsymbol{u}_0, \boldsymbol{u}_1, \boldsymbol{u}_3, \boldsymbol{u}_2).$$

(iii) If $(\boldsymbol{e}_0, \boldsymbol{e}_1, \boldsymbol{e}_2, \boldsymbol{e}_3)$ is a right-handed ON-basis (with $\boldsymbol{e}_0$ future directed), then

$$\varepsilon(\boldsymbol{e}_0, \boldsymbol{e}_1, \boldsymbol{e}_2, \boldsymbol{e}_3) = 1. \tag{4.22}$$

Conditions (i) and (ii) fix ε up to a factor. This is because the component $\varepsilon_{0123} = \varepsilon(\boldsymbol{e}_0, \boldsymbol{e}_1, \boldsymbol{e}_2, \boldsymbol{e}_3)$ of ε determines any other component $\varepsilon_{\alpha\beta\gamma\delta} = \varepsilon(\boldsymbol{e}_\alpha, \boldsymbol{e}_\beta, \boldsymbol{e}_\gamma, \boldsymbol{e}_\delta)$ by virtue of the antisymmetry condition (ii). Condition (iii) then determines ε uniquely by fixing the component ε_{0123}. At first sight it may seem that this condition implies that ε is basis-dependent. That this is not the case can be seen as follows.

From the results of the previous section we know that a basis is ON if

$$\boldsymbol{e}_\alpha\cdot\boldsymbol{e}_\beta = \eta_{\alpha\beta}. \tag{4.23}$$

Thus, if $\mathbf{e}'_\alpha = L^\gamma_\alpha \mathbf{e}_\gamma$ is any other ON-basis, we have $(L^\gamma_\alpha \mathbf{e}_\gamma)\cdot(L^\lambda_\beta \mathbf{e}_\lambda) = \eta_{\alpha\beta}$, which gives

$$L^\gamma_\alpha L^\lambda_\beta \eta_{\gamma\lambda} = \eta_{\alpha\beta} \tag{4.24}$$

and implies that $(\det L)^2 = 1$ and hence $\det L = \pm 1$. The mapping $\mathbf{e}_\alpha \to L^\gamma_\alpha \mathbf{e}_\gamma$ is called a **Lorenz transformation**. Since we are dealing only with right-handed ON bases, with $\mathbf{e}_0$ future-pointing, we see that L^γ_α can be obtained from the identity transformation δ^γ_α in a continuous fashion. Thus, since $\det \delta = 1$, we must have $\det L = 1$. Here δ^γ_α, which is called the **Kronecker delta symbol**, is defined by $\delta^\gamma_\alpha = 1$ if $\alpha = \gamma$ and $\delta^\gamma_\alpha = 0$ if $\alpha \neq \gamma$. Now, using the properties of determinants, we have $\varepsilon(\mathbf{e}'_0, \mathbf{e}'_1, \mathbf{e}'_2, \mathbf{e}'_3) = (\det L)\varepsilon(\mathbf{e}_0, \mathbf{e}_1, \mathbf{e}_2, \mathbf{e}_3)$, and hence condition (iii) holds for all right-handed ON bases with $\mathbf{e}_0$ future-pointing. The mapping ε, which is called the **Levi–Cevita tensor**, is thus a basis-independent, geometrical object. As we shall see, it has many applications in spacetime physics. Its property that we shall use most frequently is this: if one of the vectors of $(\boldsymbol{u}_0, \boldsymbol{u}_1, \boldsymbol{u}_2, \boldsymbol{u}_3)$ is a linear combination of the others, then $\varepsilon(\boldsymbol{u}_0, \boldsymbol{u}_1, \boldsymbol{u}_2, \boldsymbol{u}_3) = 0$. This follows directly from linearity and antisymmetry, which imply, for example, that $\varepsilon(\boldsymbol{u}_0, \boldsymbol{u}_1, \boldsymbol{u}_2, \boldsymbol{u}_2) = 0$.

If $\boldsymbol{v}$ is a four-velocity vector and $\boldsymbol{a}$, $\boldsymbol{b}$, and $\boldsymbol{c}$ lie in $\boldsymbol{v}^\perp$, then

$$\varepsilon(\boldsymbol{v}, \boldsymbol{a}, \boldsymbol{b}, \boldsymbol{c}) = \boldsymbol{a}\cdot(\boldsymbol{b}\times\boldsymbol{c}) \tag{4.25}$$

where $\times$ denotes the usual cross product on the Euclidean three-space $\boldsymbol{v}^\perp$. This can be easily checked by using an ON-basis such that $\boldsymbol{v} = \mathbf{e}_0$. Thus, up to a sign, we have

$$\boldsymbol{a}\cdot\boldsymbol{A} = V = \varepsilon(\boldsymbol{v}, \boldsymbol{a}, \boldsymbol{b}, \boldsymbol{c}), \tag{4.26}$$

where $\boldsymbol{A}$ is the vector area spanned by $\boldsymbol{b}$ and $\boldsymbol{c}$, and V is the volume spanned by $\boldsymbol{a}$, $\boldsymbol{b}$, and $\boldsymbol{c}$.

If $\mathbf{e}_\alpha$ is a basis such that $\boldsymbol{v} = \mathbf{e}_0$ and $\boldsymbol{n} = \mathbf{e}_1$, then, in terms of the corresponding coordinate system, we have

$$\boldsymbol{n}\cdot\delta\boldsymbol{A} = \varepsilon(\mathbf{e}_0, \mathbf{e}_1, \delta x^2\mathbf{e}_2, \delta x^3\mathbf{e}_3) = \varepsilon_{0123}\,\delta x^2\,\delta x^3, \tag{4.27}$$

$$\delta V = \varepsilon(\mathbf{e}_0, \delta x^1\,\mathbf{e}_1, \delta x^2\,\mathbf{e}_2, \delta x^3\,\mathbf{e}_3) = \varepsilon_{0123}\,\delta x^1\,\delta x^2\,\delta x^3, \tag{4.28}$$

where $\delta\boldsymbol{A}$ is the vector area spanned by the coordinate displacements δx^2 and δx^3, and δV is the volume spanned by δx^1, δx^2, and δx^3.

After that little mathematical interlude, let us return to our intrepid astronauts, Peter and Pat, who now find themselves in an interstellar dust cloud. If Peter is at rest with respect to the dust particles, we say that he is a **comoving observer**. In this case the dust particles will appear to Peter like suspended snowflakes on a still winter's night, but for any other state of motion it will seem as if he were in a blizzard. If they are massless particles such as photons, then Peter can never attain the status of a comoving observer – it will always seem as if he were in a blizzard. Peter wishes to measure the density of dust particles in a small region of volume V about a point O on his world line, where, for convenience, he sets his clock to zero. If, at $t = 0$, he finds there are N particles in V, he will say that the density is N/V. Let us now suppose that Peter's four-velocity is $\boldsymbol{v}$ and that he carries a box whose sides have displacement vectors $\boldsymbol{a}$, $\boldsymbol{b}$, and $\boldsymbol{c}$ belonging to $\boldsymbol{v}^{\perp}$ and whose volume is thus $V = \varepsilon(\boldsymbol{v}, \boldsymbol{a}, \boldsymbol{b}, \boldsymbol{c})$. If N dust world lines intersect V, Peter will say that there are N particles in V at $t = 0$ and hence that the density is $\rho(\boldsymbol{v}) = N/V$. A unique vector $\boldsymbol{j}$, called the **current density** of the dust distribution, can be determined by demanding that it point along the dust world lines and that

$$\rho(\boldsymbol{v}) = \boldsymbol{j} \cdot \boldsymbol{v}. \tag{4.29}$$

We shall now show that this equation is true for *any* four-velocity vector $\boldsymbol{v}$, where $\rho(\boldsymbol{v})$ is the corresponding particle density.

Since $\boldsymbol{v}$, $\boldsymbol{a}$, $\boldsymbol{b}$, and $\boldsymbol{c}$ form a basis, we may write

$$\boldsymbol{j} = X\boldsymbol{v} + A\boldsymbol{a} + B\boldsymbol{b} + C\boldsymbol{c},$$

where, since $\boldsymbol{a}$, $\boldsymbol{b}$, and $\boldsymbol{c}$ belong to $\boldsymbol{v}^{\perp}$, we have $X = \boldsymbol{j} \cdot \boldsymbol{v}$. Thus

$$\varepsilon(\boldsymbol{j}, \boldsymbol{a}, \boldsymbol{b}, \boldsymbol{c}) = \boldsymbol{j} \cdot \boldsymbol{v}\, \varepsilon(\boldsymbol{v}, \boldsymbol{a}, \boldsymbol{b}, \boldsymbol{c}) = \boldsymbol{j} \cdot \boldsymbol{v}\; V = \rho(\boldsymbol{v}) V = N.$$

Let us now suppose that Pat's world line intersects Peter's at O and that she has four-velocity $\boldsymbol{v}'$ and carries a box of volume $V' = \varepsilon(\boldsymbol{v}', \boldsymbol{a}', \boldsymbol{b}', \boldsymbol{c}')$, where $\boldsymbol{a}'$, $\boldsymbol{b}'$, and $\boldsymbol{c}'$ belong to $\boldsymbol{v}'^{\perp}$. If N' dust world lines intersect V', then she will say that the density is $\rho(\boldsymbol{v}') = N'/V'$. Furthermore, by the same argument as above, we have

$$\varepsilon(\boldsymbol{j}, \boldsymbol{a}', \boldsymbol{b}', \boldsymbol{c}') = \boldsymbol{j} \cdot \boldsymbol{v}' V'.$$

Let us now choose $\boldsymbol{a}'$, $\boldsymbol{b}'$, and $\boldsymbol{c}'$ such that

$$\boldsymbol{a}' = \boldsymbol{a} + \alpha \boldsymbol{j}, \qquad \boldsymbol{b}' = \boldsymbol{b} + \beta \boldsymbol{j}, \qquad \boldsymbol{c}' = \boldsymbol{c} + \gamma \boldsymbol{j},$$

in which case Pat's box will capture the same number of particles as Peter's box ($N' = N$) (see Fig. 4.7) and

$$\varepsilon(\boldsymbol{j}, \boldsymbol{a}', \boldsymbol{b}', \boldsymbol{c}') = \varepsilon(\boldsymbol{j}, \boldsymbol{a}, \boldsymbol{b}, \boldsymbol{c}).$$

Figure 4.7. If $\boldsymbol{a}' = \boldsymbol{a} + \alpha \boldsymbol{j}$, $\boldsymbol{b}' = \boldsymbol{b} + \beta \boldsymbol{j}$, $\boldsymbol{c}' = \boldsymbol{c} + \gamma \boldsymbol{j}$, then the box spanned by $\boldsymbol{a}$, $\boldsymbol{b}$, and $\boldsymbol{c}$ will capture the same number of particles as that spanned by $\boldsymbol{a}'$, $\boldsymbol{b}'$, and $\boldsymbol{c}'$.

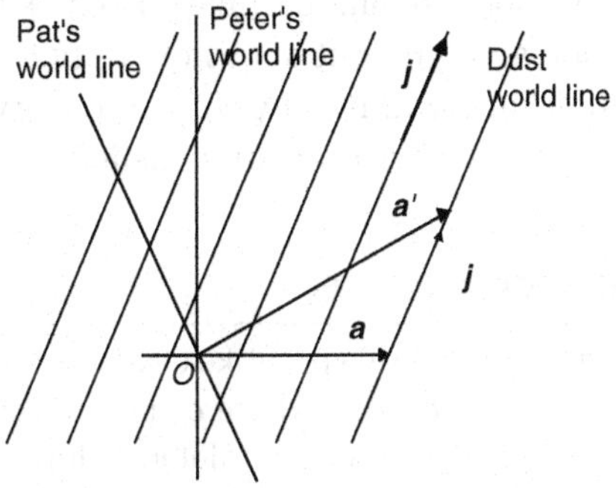

Thus

$$j \cdot v'V' = \varepsilon(j, a', b', c') = \varepsilon(j, a, b, c) = N = N',$$

which shows that the density according to Pat is given by $\rho(v') = j \cdot v'$. We have thus proved:

Theorem 4.4 *For any uniform dust distribution there exists a vector, j, called the current density, such that $\rho(v) = j \cdot v$, where $\rho(v)$ is the particle number density according to an observer with four-velocity v.*

For massive dust particles the current will have the form $j = \rho_0 v_0$, where v_0 is the four-velocity vector of a comoving observer. In this case $\rho(v_0) = j \cdot v_0 = \rho_0$, and hence ρ_0 is the density according to a comoving observer. Note that, by inequality (4.7), $\rho(v) \geq \rho_0$ and hence ρ_0 is the least possible density. However, for massless particles where j is null, $\rho(v)$ can assume any positive value. In particular, by choosing v suitably, $\rho(v)$ can be made arbitrarily small.

If each of the particles carries a electric charge e, then the **charge density** according to an observer with four-velocity v will be $\rho_e(v) = e\rho(v)$, and the **electric current** is defined to be $j_e = ej$.

So far we have considered uniform particle distributions where, at least locally, all particles have the same four-velocity. Physically, this corresponds to something like a blizzard where all the snowflakes have the same speed and move in the same direction. However, for something like a dilute gas in thermal equilibrium, the gas molecules do not all have the same speed or direction but form a higgledy-piggledy, nonuniform particle distribution. Nevertheless, such a distribution can be thought of as a superposition of uniform distributions, and its current can be written

$$j = \sum_i \underset{(i)}{j}$$

where $\underset{(i)}{j}$ are the uniform currents of its component parts. However, such a description breaks down if we allow the particles to coalesce or bifurcate, because the number of particles will then vary with time, and in this case the total particle current j will not be well defined. Fortunately, such an argument does not apply to the electric current of a nonuniform distribution, since, by charge conservation, the total charge of a collection of particles will remain constant even though their number may vary.

EXERCISES

4.1 A natural spacelike ON-basis for an observer sitting at the North Pole is (e_1, e_2, e_3), where e_3 points vertically upwards (toward the polestar). This can be completed to form a four-basis

$$(e_0, e_1, e_2, e_3),$$

where $\boldsymbol{e}_0 = \boldsymbol{v}$ is the observer's four-velocity. A null ray with tangent vector $\boldsymbol{p}$ will appear to come from the polestar if $\boldsymbol{p} \cdot \boldsymbol{e}_1 = \boldsymbol{p} \cdot \boldsymbol{e}_2 = 0$. Use this fact to show how the number density of the particle distribution of photons arriving from the pole star changes if the observer moves (a) horizontally and (b) vertically upwards, with speed W.

4.2 Derive the Doppler-shift formula for observers whose world lines do not intersect. This should be expressed as a function of proper time.

4.3 Describe how the distribution of stars in the sky would appear to change if you got into a spaceship and traveled with a high speed toward the polestar.

4.4 To an observer on earth the full moon appears as a disk. How would it appear to an observer flying past in a spaceship?

5

Energy

As anyone who has paid an electricity bill knows, energy is a very real physical quantity; but unlike other expensive commodities such as books, records, and bottles of wine, it has a universal and all-pervading character. But what actually is energy? We shall attempt to answer this question by simply listing its defining properties. We must, however, be careful to list enough properties so as to capture the notion of energy as is actually used in fundamental physics, and also to distinguish it from other conserved quantities such as electric charge. We must, of course, not list too many properties, or properties that are too rigid, as this may lead to a trivial or nonexistent quantity. We do not want, for example, to end up with a quantity that turns out to be identically zero.

Energy is a measurable quantity possessed by *all* physical objects, which is always found to be *strictly positive*. Energy comes, of course, in many forms, ranging from pure radiative energy, which tends to be its most useful but most dangerous form, to its most benign form, which is safely locked away, like a genie in a lamp, in all inert matter. Even the paperweight in front of me contains energy (quite a considerable amount) but, fortunately for me, this is in its most benign form. I could, in principle, if I so wished, convert this energy into its radiative form by bringing my paper weight into contact with an otherwise identical one made of antimatter, and this would result in a spectacular firework display of pure radiative energy. (Such experiments are, however, not recommended, as they can cause quite a mess.) It is this universal character of energy that distinguishes it from all other conserved quantities and that makes it so interesting from the point of view of spacetime physics. Not all physical objects have, for example, a strictly positive electric charge, and even if they have a nonzero charge, it can be negative. No physicist would even think of measuring the electric charge of a photon, because a photon, by its very nature, is chargeless. He would, on the other hand, not think it unreasonable to measure the energy of a photon and would be surprised if the result turned out to be negative. He would be equally surprised, but more pleasantly so, if he received an electricity bill demanding a negative sum of money.

It would be wrong to say that all objects contain, in some absolute sense, a definite amount of energy, because the result of an energy measurement depends very much on the relative motion between the observer and the system. Electric charge, on the other hand, is independent of this motion. The energy (frequency) of a photon, for example, can take any positive value whatsoever, depending on the four-velocity of the observer [see equation (4.12)].

A crucial property of energy is, of course, that it is conserved. By this we mean that if we have two isolated objects with total energies E_1 and E_2 which we combine to make one isolated object, then its total energy will be $E = E_1 + E_2$. If this turns out not to be true, then either the quantity measured is not energy or the *total* energy of the object has not been taken into account. The history of physics is full of accounts of missing energy – energy that, at first sight, seems to have simply vanished, thus violating the conservation law. However, this has always been found to be due to some hitherto unexpected new form of energy that had not been taken into account.

There is one final property of energy, which follows essentially from the fact that it is a *linear* function of the observer's four-velocity in the particular case of a photon [see equation (4.12)]. This, together with the fact that energy is conserved, implies that the energy of *any* object must be a linear function of the observer's four-velocity. If that were not the case, then interactions that produce or absorb photons would lead to inconsistencies.

5.1 Energy and Four-Momentum

We denote the total energy of a physical system, as measured by an observer with four-velocity $\boldsymbol{v}$, by $E(\boldsymbol{v})$. Since one of the defining properties of energy is that it is linear in $\boldsymbol{v}$, we have $E(\boldsymbol{v}) = \boldsymbol{v} \cdot \boldsymbol{p}$ for some unique vector $\boldsymbol{p}$, which is called the **four-momentum** of the system. Since $E(\boldsymbol{v})$ is strictly positive, $\boldsymbol{p}$ must be future-pointing. Unlike $E(\boldsymbol{v})$, which depends on $\boldsymbol{v}$, the four-momentum vector depends only on the physical system itself and is therefore an intrinsic, observer-independent property of it. By linearity and conservation of energy we see that if $\boldsymbol{p}_1$ and $\boldsymbol{p}_2$ are the four-momenta of two isolated systems that are brought together to form a combined isolated system, then the four-momentum of the combined system is $\boldsymbol{p} = \boldsymbol{p}_1 + \boldsymbol{p}_2$.

If $\boldsymbol{p}$ is timelike, then there will exist a four-velocity vector $\boldsymbol{w}$ and a positive number m such that $\boldsymbol{p} = m\boldsymbol{w}$. Since, by inequality (4.7), $\boldsymbol{v} \cdot \boldsymbol{w} \geq 1$ (equality when $\boldsymbol{v} = \boldsymbol{w}$), we see that m is the least possible energy of the system. Note that $\boldsymbol{w} \cdot \boldsymbol{p} = m$ and $\boldsymbol{p} \cdot \boldsymbol{p} = m^2$. The number m, which is an intrinsic property of the system, is called its **rest mass**. If $\boldsymbol{p}$ is null, then $\boldsymbol{p} \cdot \boldsymbol{p} = 0$ and we say that the system's rest mass is zero.

If our system is a single particle, then, by the symmetry of the situation, $\boldsymbol{w}$ will coincide with the particle's four-velocity. In this case the energy assumes its minimum value when the observer is at rest relative to the particle ($\boldsymbol{w} = \boldsymbol{v}$). Its rest mass is thus the *nonkinetic* part of its energy. Using equation (4.6) we see that

$$E(\boldsymbol{v}) = m\boldsymbol{v} \cdot \boldsymbol{w} = \frac{m}{\sqrt{1-v^2}} = m + \frac{mv^2}{2} + O(v^4), \tag{5.1}$$

where v is the relative speed between the observer and the particle. This shows that, for speeds small compared to that of light (recall that we are using units such that $c = 1$), the *kinetic* part of the energy is compatible with that of Newtonian physics.

By writing $\boldsymbol{w} = \gamma(\boldsymbol{v} + \boldsymbol{V})$ where $\gamma = \boldsymbol{v} \cdot \boldsymbol{w}$ and $\boldsymbol{V} \in \boldsymbol{v}^{\perp}$ we have

$$\boldsymbol{p} = m\gamma\boldsymbol{v} + m\gamma\boldsymbol{V} = E(\boldsymbol{v})\boldsymbol{v} + E(\boldsymbol{v})\boldsymbol{V}.$$

The vector $\boldsymbol{P} = E(\boldsymbol{v})\boldsymbol{V}$ is called the system's total **three-momentum** relative to the observer. Unlike $\boldsymbol{p}$, which is an observer-independent quantity, $\boldsymbol{P}$ depends on the observer's four-velocity. It is, however, a conserved quantity and can be equated with Newtonian momentum. For this reason, we sometimes write

$$E(\boldsymbol{v}) = M(\boldsymbol{v}) = \frac{m}{\sqrt{1-v^2}} \tag{5.2}$$

and $\boldsymbol{P} = M(\boldsymbol{v})\boldsymbol{V}$, and interpret $M(\boldsymbol{v})$ as the total mass of the system according to an observer with four-velocity $\boldsymbol{v}$. The conservation of $\boldsymbol{p}$ or, equivalently, the conservation of E for *all* inertial observers, implies the conservation of both M (total mass–energy) and $\boldsymbol{P}$ (total three-momentum) for any particular observer.

Example 5.1 A photon with four-momentum $\boldsymbol{p}$ scatters off an electron with four-momentum $m\boldsymbol{v}$ so as to produce a photon of four-momentum $\boldsymbol{p}'$ and an electron of four-momentum $m\boldsymbol{v}'$ (Fig. 5.1). Here m is the rest mass of an electron and $\boldsymbol{v}$ and $\boldsymbol{v}'$ are four-velocities. Let us find a relation (the Compton scattering formula) between the scattering angle between the photons and their energies, according to an observer with four-momentum $\boldsymbol{v}$, that is, an observer at rest relative to the initial electron.

Figure 5.1. Compton scattering.

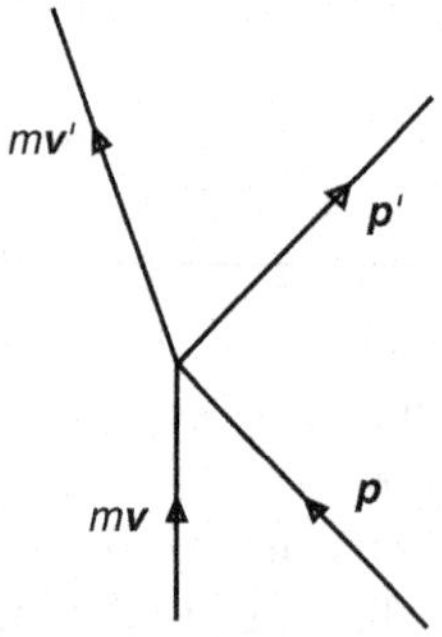

Writing $\boldsymbol{p} = E(\boldsymbol{v} + \boldsymbol{e})$ and $\boldsymbol{p}' = E'(\boldsymbol{v} + \boldsymbol{e}')$, where $\boldsymbol{e}$ and $\boldsymbol{e}'$ are unit, spacelike vectors orthogonal to $\boldsymbol{v}$, we have $E = \boldsymbol{p} \cdot \boldsymbol{v}$, $E' = \boldsymbol{p}' \cdot \boldsymbol{v}$, and $\boldsymbol{p} \cdot \boldsymbol{p}' = EE'(1 + \boldsymbol{e} \cdot \boldsymbol{e}')$. E and E' are thus the energies to the two photons. Since $\boldsymbol{e}$ and $\boldsymbol{e}'$ correspond to their directions, we have

$\boldsymbol{e} \cdot \boldsymbol{e}' = -\cos\theta$, where θ is the scattering angle. Thus

$$\boldsymbol{p} \cdot \boldsymbol{p}' = EE'(1 - \cos\theta). \tag{5.3}$$

By conservation of four-momentum, $m\boldsymbol{v} + \boldsymbol{p} - \boldsymbol{p}' = m\boldsymbol{v}'$, and since $\boldsymbol{v}' \cdot \boldsymbol{v}' = 1$, we get

$$mE = mE' + \boldsymbol{p} \cdot \boldsymbol{p}'. \tag{5.4}$$

Equations 5.3 and 5.4 now give

$$m(E - E') = EE'(1 - \cos\theta), \tag{5.5}$$

which is the celebrated formula for Compton scattering.

5.2 The Energy–Momentum Tensor

Given a quasi-continuous distribution of dust particles, each with four-momentum $\boldsymbol{p}$, the **four-momentum density**, $\boldsymbol{\rho}(\boldsymbol{v})$, according to an observer with four-velocity $\boldsymbol{v}$ is given by

$$\boldsymbol{\rho}(\boldsymbol{v}) = \boldsymbol{p}\rho(\boldsymbol{v}) = \boldsymbol{p}(\boldsymbol{j} \cdot \boldsymbol{v}),$$

where $\rho(\boldsymbol{v})$ is the particle-number density and $\boldsymbol{j}$ is its current. The total four-momentum and total energy contained in a small region of $\boldsymbol{v}^{\perp}$ with volume V will thus be given by $\boldsymbol{p}_V = \boldsymbol{\rho}(\boldsymbol{v})V$ and $E_V = [\boldsymbol{\rho}(\boldsymbol{v}) \cdot \boldsymbol{v}]V = (\boldsymbol{j} \cdot \boldsymbol{v})(\boldsymbol{p} \cdot \boldsymbol{v})V$. The quadratic form defined by

$$T(\boldsymbol{a}, \boldsymbol{b}) = (\boldsymbol{j} \cdot \boldsymbol{a})(\boldsymbol{p} \cdot \boldsymbol{b}) \tag{5.6}$$

is called the **energy–momentum tensor** of the distribution. Since both $\boldsymbol{j}$ and $\boldsymbol{p}$ both point in the same direction (i.e. along the world lines of the dust particles) and are both future-pointing, we see that T is symmetric in that $T(\boldsymbol{a}, \boldsymbol{b}) = T(\boldsymbol{b}, \boldsymbol{a})$ and that

$$T(\boldsymbol{w}, \boldsymbol{w}) \geq 0 \tag{5.7}$$

for any *timelike* vector $\boldsymbol{w}$. This is sometimes referred to as the **weak energy condition**. Furthermore, for massless particles, we have

$$\operatorname{tr} T = 0, \tag{5.8}$$

where $\operatorname{tr} T$, the **trace** or **contraction** of T, is defined by $\operatorname{tr} T = \boldsymbol{j} \cdot \boldsymbol{p}$. For massive dust particles with four-velocity $\boldsymbol{v}$ we have $\boldsymbol{j} = \rho_0 \boldsymbol{v}$ and $\boldsymbol{p} = m\boldsymbol{v}$ and thus

$$T(\boldsymbol{a}, \boldsymbol{b}) = m\rho_0(\boldsymbol{v} \cdot \boldsymbol{a})(\boldsymbol{v} \cdot \boldsymbol{b}), \tag{5.9}$$

$$\operatorname{tr} T = m\rho_0. \tag{5.10}$$

In component form this gives $T^{\alpha\beta} = m\rho_0 v^\alpha v^\beta$ and $T^\alpha{}_\alpha = m\rho_0$. These two equations imply that the tensor given by

$$T'(\boldsymbol{a}, \boldsymbol{b}) = T(\boldsymbol{a}, \boldsymbol{b}) - \tfrac{1}{2}(\boldsymbol{a} \cdot \boldsymbol{b}) \operatorname{tr} T, \tag{5.11}$$

satisfies

$$T'(\boldsymbol{w}, \boldsymbol{w}) = T(\boldsymbol{w}, \boldsymbol{w}) - \tfrac{1}{2} \operatorname{tr} T \geq 0 \tag{5.12}$$

for any four-velocity vector $\boldsymbol{w}$. This inequality, which is known as the **strong energy condition**, will prove important when we come to consider gravity.

As we saw in Section 4.3, a nonuniform, higgledy-piggledy distribution where the particles are not restricted to have the same four-velocity can be thought of as a superposition of uniform particle distributions. For such a distribution, for example, a gas in thermal equilibrium, we may thus write the energy–momentum tensor as

$$T = \sum_i \underset{i}{T}$$

where $\underset{i}{T}$ are the energy–momentum tensors of its uniform components. Even if we do not have particle-number conservation, the conservation of four-momentum implies that the energy–momentum tensor for a nonuniform distribution is still well defined. Note that the weak and strong energy conditions still hold, and that for massless particles we have

$$\operatorname{tr} T = \sum_i \operatorname{tr} \underset{i}{T} = 0. \tag{5.13}$$

For a nonrelativistic gas where the speed of the gas molecules is small compared to that of light, we may write $\underset{i}{\boldsymbol{v}} \approx \boldsymbol{v}$, where $\boldsymbol{v}$ is, for example, the four-velocity of the gas container, and hence

$$T(\boldsymbol{a}, \boldsymbol{b}) \approx \sum_i m\underset{i}{\rho_0}(\boldsymbol{a} \cdot \boldsymbol{v})(\boldsymbol{b} \cdot \boldsymbol{v}) = \rho(\boldsymbol{a} \cdot \boldsymbol{v})(\boldsymbol{b} \cdot \boldsymbol{v}), \tag{5.14}$$

where $\rho = \sum_i m\underset{i}{\rho_0}$ is the average mass density of the gas.

5.3 General States of Matter

So far we have considered only "billiard-ball" systems consisting of a quasi-continuous collection of particles. If the particles do not interact (e.g. a dilute gas) then the particle-number current $\boldsymbol{j}$ is well defined, but it ceases to be meaningful if we allow collisions that change the number of particles. Nevertheless, due to charge and energy conservation, the electric current $\boldsymbol{j}_e$ and the energy–momentum tensor T remain well defined.

We have, however, neglected fields such as electromagnetism, which also carry energy and which mediate the interaction between particles. In this section we shall rectify this omission by considering general states of matter.

Given some matter distribution, for example, an electromagnetic field, a small region of $v^{\perp}$ of volume V will capture a certain amount of four-momentum, $\boldsymbol{p}$ say. Assuming V is small, we make the following two physical assumptions:

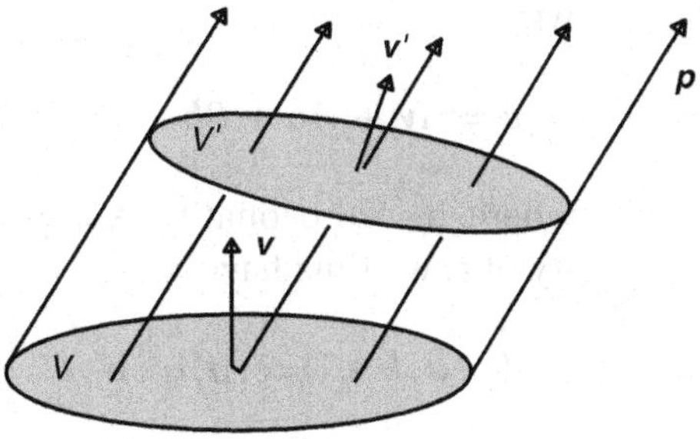

Figure 5.2. Regions V and V' capture the same amount of four-momentum.

- $\boldsymbol{p}$ is proportional to V, that is, $\boldsymbol{p} = \rho(\boldsymbol{v})V$ for some vector $\rho(\boldsymbol{v})$ depending on $\boldsymbol{v}$. We call $\rho(\boldsymbol{v})$ the four-momentum density.
- Given a small four-dimensional region with floor V and ceiling V' and whose walls are parallel to $\boldsymbol{p}$, then V and V' capture the same amount of four-momentum (see Fig. 5.2), namely,

$$\boldsymbol{p} = \rho(\boldsymbol{v})V = \rho(\boldsymbol{v}')V'.$$

The four-momentum density $\rho(\boldsymbol{v})$ gives a physically defined mapping that takes a *four-velocity* vector $\boldsymbol{v}$ and gives a new vector $\rho(\boldsymbol{v})$. We cannot at this stage assume that this mapping is defined on vectors other than four-velocity vectors or that it is linear. Note that, for a "billiard-ball" system, we have $T(\boldsymbol{v}, \boldsymbol{v}) = \rho(\boldsymbol{v})\cdot\boldsymbol{v}$. Our first main result concerning $\rho(\boldsymbol{v})$ is this:

Theorem 5.1

$$\rho(\boldsymbol{v})\cdot\boldsymbol{v}' = \rho(\boldsymbol{v}')\cdot\boldsymbol{v}$$

for any two four-velocity vectors $\boldsymbol{v}$ *and* $\boldsymbol{v}'$.

Proof If $\boldsymbol{a}, \boldsymbol{b}, \boldsymbol{c} \in \boldsymbol{v}^{\perp}$ and $\boldsymbol{a}', \boldsymbol{b}', \boldsymbol{c}' \in \boldsymbol{v}'^{\perp}$, then the corresponding volumes are $V = \varepsilon(\boldsymbol{v}, \boldsymbol{a}, \boldsymbol{b}, \boldsymbol{c})$ and $V' = \varepsilon(\boldsymbol{v}', \boldsymbol{a}', \boldsymbol{b}', \boldsymbol{c}')$. In general $\boldsymbol{p} = \rho(\boldsymbol{v})V \neq \boldsymbol{p}' = \rho(\boldsymbol{v}')V'$, but if $\boldsymbol{a}' = \alpha\boldsymbol{p} + \boldsymbol{a}$, $\boldsymbol{b}' = \beta\boldsymbol{p} + \boldsymbol{b}$, and $\boldsymbol{c}' = \gamma\boldsymbol{p} + \boldsymbol{c}$, then V and V' will capture the same amount of four-momentum, and hence $\boldsymbol{p} = \boldsymbol{p}'$ (see Fig. 5.3). Since $(\boldsymbol{v}, \boldsymbol{a}, \boldsymbol{b}, \boldsymbol{c})$ and $(\boldsymbol{v}', \boldsymbol{a}', \boldsymbol{b}', \boldsymbol{c}')$ form bases, we may

Figure 5.3. The box spanned by $\boldsymbol{a}$, $\boldsymbol{b}$, and $\boldsymbol{c}$ captures the same four-momentum as that spanned by $\boldsymbol{a}' = \alpha\boldsymbol{p} + \boldsymbol{a}$, $\boldsymbol{b}' = \beta\boldsymbol{p} + \boldsymbol{b}$, and $\boldsymbol{c}' = \gamma\boldsymbol{p} + \boldsymbol{c}$.

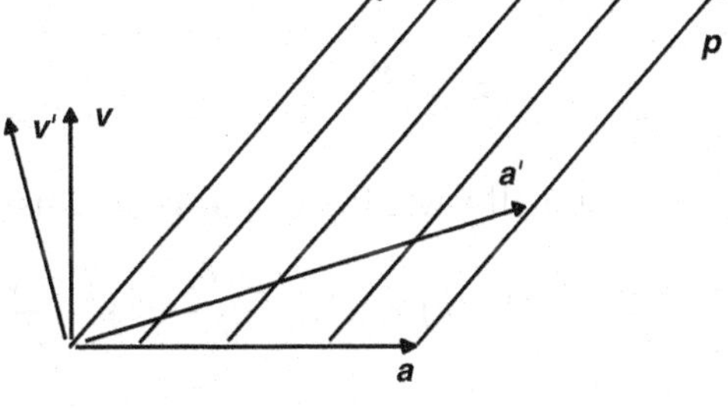

write

$$p = Xv + Aa + Bb + Cc = X'v' + A'a' + B'b' + C'c',$$

where, by orthogonality, $X = p \cdot v$ and $X' = p \cdot v'$. Using the antisymmetry of ε, we thus have

$$\varepsilon(p, a, b, c) = \varepsilon(p, a', b', c') = p \cdot v\, V = p \cdot v' V',$$

which, since $p = p'$, can be written $p' \cdot v\, V = p \cdot v' V'$ or $\rho(v') \cdot v\, VV' = \rho(v) \cdot v' VV'$. □

Our next theorem shows that $\rho(v) \cdot v$ defines a quadratic from T, namely the energy–momentum tensor.

Theorem 5.2 *There exists a unique quadratic form T such that*

$$T(v, v) = \rho(v) \cdot v.$$

Proof Let $(v_\alpha, a_\alpha, b_\alpha, c_\alpha)$ $(\alpha = 0, 1, 2, 3)$ be bases such that $a_\alpha, b_\alpha, c_\alpha \in v_\alpha^\perp$ and such that (v_0, v_1, v_2, v_3) is a basis of four-velocity vectors. We define T by

$$T(v_\alpha, v_\beta) = \rho(v_\alpha) \cdot v_\beta$$

and wish to show that $T(v, v) = \rho(v) \cdot v$ for any four-velocity vector v. Let $a, b, c \in v^\perp$, and let p be the total four-momentum contained in $V = \varepsilon(v, a, b, c)$. For convenience we let $V = 1$. If $a_\alpha = a + \alpha_\alpha p$, $b_\alpha = b + \beta_\alpha p$, and $c_\alpha = c + \gamma_\alpha p$, then the corresponding volumes V_0, V_1, V_2, and V_3 capture the same four-momentum as V, namely p. Thus $p = \rho(v_\alpha) V_\alpha$ and hence

$$T(v_\alpha, v_\beta) = \rho(v_\alpha) \cdot v_\beta = \frac{p \cdot v_\beta}{V_\alpha}. \tag{5.15}$$

Furthermore, from the previous proof we see that

$$\varepsilon(p, a, b, c) = \varepsilon(p, a_\alpha, b_\alpha, c_\alpha) = p \cdot v_\alpha V_\alpha = p \cdot v,$$

or

$$p \cdot v_\alpha = \frac{p \cdot v}{V_\alpha}. \tag{5.16}$$

Finally, writing $v = x^\alpha v_\alpha$ and using equations (5.15) and (5.16), we have

$$T(v, v) = x^\alpha x^\beta \frac{p \cdot v_\beta}{V_\alpha} = x^\alpha \frac{p \cdot v}{V_\alpha} = x^a (p \cdot v_\alpha) = p \cdot v = \rho(v) \cdot v.$$

□

Note that the energy contained in V is given by

$$E = T(\boldsymbol{v}, \boldsymbol{v})V = \rho(\boldsymbol{v}) \cdot \boldsymbol{v} V \geq 0. \tag{5.17}$$

Positivity of energy thus implies that T satisfies the weak energy condition. A simple extension of the above proof also shows that

$$T(\boldsymbol{v}, \boldsymbol{v}') = \rho(\boldsymbol{v}) \cdot \boldsymbol{v}' \tag{5.18}$$

for any two four-velocity vectors, $\boldsymbol{v}$ and $\boldsymbol{v}'$. Since, for positive energy, $\rho(\boldsymbol{v})$ is future-pointing, we have

$$T(\boldsymbol{v}, \boldsymbol{v}') \geq 0. \tag{5.19}$$

Positivity of energy alone does not, however, imply the strong energy condition:

$$T(\boldsymbol{v}, \boldsymbol{v}) - \tfrac{1}{2} \operatorname{tr} T \geq 0.$$

This is unfortunate, since it is precisely this energy condition that we shall employ when we come to consider gravity. In fact, we shall show that this condition is intimately related to the attractiveness of gravity. It is, however, a remarkable fact that all physically reasonable forms of matter seem to satisfy it, and, as we have seen, it is certainly satisfied by "billiard-ball" systems. For a full discussion of the strong energy condition see Hawking and Ellis (1973).

Like any other bilinear mapping, T defines a linear mapping $\boldsymbol{a} \to T(\boldsymbol{a})$ by $T(\boldsymbol{a}, \boldsymbol{b}) = T(\boldsymbol{a}) \cdot \boldsymbol{b}$ for all $\boldsymbol{b}$. This mapping is just another form of the energy–momentum tensor, so, with a slight abuse of notation, we represent it by the same symbol, namely T. Note that $T(\boldsymbol{v}) = \rho(\boldsymbol{v})$, where $\boldsymbol{v}$ is a four-velocity vector.

5.4 Perfect Fluids

A **perfect fluid** is a type of matter distribution that does not single out a preferred direction in space according to some (comoving) observer who is at rest with respect to the general flow of the fluid. For example, to an underwater swimmer at rest in a calm swimming pool, all directions will be alike – assuming, of course, that the swimming pool is infinitely large and is in interstellar space. Clearly, matter in a model universe satisfying the cosmological principle must be described by a perfect fluid.

The energy–momentum tensor (in linear-mapping form) defines at most four directions, namely the directions given by its eigenvectors. For the energy–momentum of a perfect fluid, one of these must be the four-velocity $\boldsymbol{v}$ of a comoving observer [i.e. $T(\boldsymbol{v}) = \rho\boldsymbol{v}$], and the other three must lie in $\boldsymbol{v}^{\perp}$. (Why?) Furthermore, by isotropy, they must be degenerate and have eigenspace $\boldsymbol{v}^{\perp}$ and some eigenvalue p given by $T(\boldsymbol{e}) = p\boldsymbol{e}$ for all

$\boldsymbol{e} \in \boldsymbol{v}^{\perp}$. Note that the eigenvalue ρ of $\boldsymbol{v}$ is given by $\rho = T(\boldsymbol{v})\cdot\boldsymbol{v} = \rho(\boldsymbol{v})\cdot\boldsymbol{v}$ and is thus the energy density with respect to a comoving observer. The eigenvalue p of $\boldsymbol{e}$ is, by definition, the **pressure** of the fluid. This definition will be justified later.

By introducing a ON-basis such that $\boldsymbol{v} = \boldsymbol{e}_0$, we have $T(\boldsymbol{e}_0) = \rho\boldsymbol{e}_0$ and $T(\boldsymbol{e}_i) = p\boldsymbol{e}_i$, which shows that

$$T(\boldsymbol{a}, \boldsymbol{b}) = T(\boldsymbol{a})\cdot\boldsymbol{b} = (\rho + p)(\boldsymbol{v}\cdot\boldsymbol{a})(\boldsymbol{v}\cdot\boldsymbol{b}) - p\boldsymbol{a}\cdot\boldsymbol{b}, \tag{5.20}$$

or, in component form,

$$T_{\alpha\beta} = (\rho + p)v_\alpha v_b - p\eta_{\alpha\beta}. \tag{5.21}$$

(This can be checked by taking $\boldsymbol{a}$ and $\boldsymbol{b}$ to be basis vectors.) Since $\operatorname{tr} g = 4$ (g is the metric), we have

$$\operatorname{tr} T = \rho - 3p, \tag{5.22}$$

and hence, for a perfect photon gas where $\operatorname{tr} T = 0$ [see equation (5.13)], we have the important equation

$$\rho = 3p. \tag{5.23}$$

Using equations (5.20) and (5.22), we get

$$T(\boldsymbol{v}, \boldsymbol{v}) - \tfrac{1}{2}\operatorname{tr} T = \frac{\rho + 3p}{2}, \tag{5.24}$$

and hence the strong energy condition holds if

$$\rho \geq -3p. \tag{5.25}$$

5.5 Acceleration and the Maxwell Tensor

Consider a noninertial (accelerating) world line with position vector $\mathbf{x}(\tau)$, where τ is the *proper-time*. Its tangent vector $\dot{\mathbf{x}} = d\mathbf{x}/d\tau$ will be a four-velocity vector (why?), and hence $\dot{\mathbf{x}}\cdot\dot{\mathbf{x}} = 1$. We define the **four-acceleration** of the world line to be $\ddot{\mathbf{x}}$. Since $\dot{\mathbf{x}}\cdot\dot{\mathbf{x}} = 1$, we have $\ddot{\mathbf{x}}\cdot\dot{\mathbf{x}} = 0$. The four-acceleration is therefore always orthogonal to its corresponding four-velocity.

In terms of components we have

$$\begin{aligned}
x^\alpha &= (t, x, y, z),\\
\dot{x}^\alpha &= (\dot{t}, \dot{x}, \dot{y}, \dot{z}) = \gamma(1, x', y', z'),\\
\ddot{x}^\alpha &= \dot{\gamma}(1, x', y', z') + \gamma^2(0, x'', y'', z''),
\end{aligned}$$

where $\dot{t} = \gamma$ and $x' = dx/dt$. In vector form these equations give $\gamma = \dot{\boldsymbol{x}} \cdot \boldsymbol{e}_0$, and

$$\dot{\boldsymbol{x}} = \gamma(\boldsymbol{e}_0 + \boldsymbol{V}), \tag{5.26}$$
$$\ddot{\boldsymbol{x}} = \dot{\gamma}(\boldsymbol{e}_0 + \boldsymbol{V}) + \gamma^2 \boldsymbol{A}, \tag{5.27}$$

where $\boldsymbol{V}, \boldsymbol{A} \in \dot{\boldsymbol{x}}^{\perp}$. Here $\boldsymbol{V}$ and $\boldsymbol{A}$ are the instantaneous three-velocity and **three-acceleration** of the particle with respect to an observer with four-velocity $\boldsymbol{e}_0$. Since $\ddot{\boldsymbol{x}} \cdot \dot{\boldsymbol{x}} = 0$, these equations give

$$\dot{\gamma} = -\gamma^4 \boldsymbol{A} \cdot \boldsymbol{V}. \tag{5.28}$$

For small three-velocity we have $\gamma \approx 1$, and if we also have small three-acceleration, this equation gives $\dot{\gamma} \approx 0$. Neglecting second-order terms, equations (5.26) and (5.27) thus reduce to $\dot{\boldsymbol{x}} = \boldsymbol{e}_0 + \boldsymbol{V}$ and $\ddot{\boldsymbol{x}} = \boldsymbol{A}$.

If an observer with four-velocity $\boldsymbol{v}$ releases a charged particle from rest ($\boldsymbol{v} = \dot{\boldsymbol{x}}$) in an electromagnetic field, the particle will have an instantaneous acceleration given by

$$m\ddot{\boldsymbol{x}} = eF(\boldsymbol{v}), \tag{5.29}$$

where $F(\boldsymbol{v})$ is the electric part of the field (with respect to the observer) and m and e are the rest mass and charge of the particle. An electromagnetic field thus gives a physically defined mapping that takes a four-velocity vector $\boldsymbol{v}$ and gives an orthogonal (space-like) vector $F(\boldsymbol{v})$, that is, $\boldsymbol{v} \cdot F(\boldsymbol{v}) = 0$. If we make the physically reasonable and well-verified assumption that F is linear, then it gives a 2-form defined by

$$F(\boldsymbol{a}, \boldsymbol{b}) = \boldsymbol{a} \cdot F(\boldsymbol{b}).$$

Clearly, $F(\boldsymbol{a}, \boldsymbol{b}) = -F(\boldsymbol{b}, \boldsymbol{a})$. This 2-form is called the **Maxwell tensor** and fully describes the electromagnetic field. We recall that the components of a 2-form F are defined by $F_{\alpha\beta} = F(\boldsymbol{e}_\alpha, \boldsymbol{e}_\beta)$ and the components of the corresponding linear mapping by $F(\boldsymbol{e}_\alpha) = F_\alpha{}^\beta \boldsymbol{e}_\beta$, where $F_{\alpha\beta} = F_\alpha{}^\sigma \eta_{\sigma\beta}$.

If our observer is not at rest relative to the particle ($\boldsymbol{e}_0 = \boldsymbol{v} \neq \dot{\boldsymbol{x}}$), then (neglecting second-order terms) equations (5.26), (5.27), and (5.29) give

$$m\boldsymbol{A} = eF(\boldsymbol{e}_0 + \boldsymbol{V}) = eF(\boldsymbol{e}_0) + eF(\boldsymbol{V}), \tag{5.30}$$

or, in component form,

$$m\ddot{x}^i = e\left(F^i{}_0 + F^i{}_j \dot{x}^j\right) \tag{5.31}$$

By writing

$$F_{\alpha\beta} = \begin{pmatrix} 0 & E_1 & E_2 & E_3 \\ -E_1 & 0 & B_3 & -B_2 \\ -E_2 & -B_3 & 0 & B_1 \\ -E_3 & B_2 & -B_1 & 0 \end{pmatrix}, \tag{5.32}$$

$\boldsymbol{E} = (E_1, E_2, E_3)$, and $\boldsymbol{B} = (B_1, B_2, B_3)$ then, in ordinary three-space vector notation, equation (5.31) becomes

$$m\ddot{\mathbf{x}} = e(\boldsymbol{E} + \boldsymbol{B} \times \dot{\mathbf{x}}), \tag{5.33}$$

which shows that the $\boldsymbol{B}$ part of $F_{\alpha\beta}$ is the magnetic part of the field.

EXERCISES

5.1 An electron and a positron decay into a shower of photons. Show that the shower must contain more than one photon.

5.2 A particle of mass M is at rest with respect to an observer. If it decays into two photons, what is their energy with respect to the observer?

5.3 If it decays into a photon and a particle of mass m, what is the energy of the photon with respect to the observer?

5.4 If it decays into two particles of masses m_1 and m_2, find the energies of these two particle with respect to the observer.

5.5 A rocket has a fixed mass of fuel, which it can expel in the form of massive particles or photons. Which process will be the more efficient?

5.6 Consider all world lines connecting two given events. Which one contains the *greatest* interval of proper time?

6

Tensors

The value of a single measurement performed at some given event gives, of course, just one number – the result of the measurement. But, in general, this number will depend on many other factors apart from the mere location of the event. It may for instance depend on the state of motion of the observer and the configuration of his apparatus. A measurement of charge density will, for example, depend on the observer's four-velocity, and a measurement of electric field strength in some direction will depend on this direction as well as the observer's four-velocity. Of particular interest are measurements that depend *linearly* on a sequence of vectors describing the motion and configuration of the apparatus. If this is the case, the quantity measured is said to be a **tensor**.

6.1 Tensors at a Point

As a preliminary definition, a tensor is simply a mapping t, linear in each of its slots, that takes a sequence of vectors, $\boldsymbol{a}, \ldots, \boldsymbol{b}$, and gives a number $t(\boldsymbol{a}, \ldots, \boldsymbol{b})$. If t has N slots, we say that it has type $(0, N)$. We have already come across several examples of tensors: the metric g [type $(0, 2)$], the Levi–Civita tensor ε [type $(0, 4)$], the energy–momentum tensor T [type $(0, 2)$], and the Maxwell tensor F [type $(0, 2)$]. g and T are symmetric, and F and ε are (completely) antisymmetric, or **skew**. In terms of a basis $(\boldsymbol{e}_\alpha)$ the components of a tensor t are defined by

$$t_{\alpha\cdots\beta} = t(\boldsymbol{e}_\alpha, \ldots, \boldsymbol{e}_\beta).$$

In particular, the components of the metric are given by

$$g_{\alpha\beta} = g(\boldsymbol{e}_\alpha, \boldsymbol{e}_\beta) = \boldsymbol{e}_\alpha \cdot \boldsymbol{e}_\beta. \tag{6.1}$$

If $\boldsymbol{a} = a^\alpha \boldsymbol{e}_\alpha$ and $\boldsymbol{b} = b^\alpha \boldsymbol{e}_\alpha$ then

$$g(\boldsymbol{a}, \boldsymbol{b}) = \boldsymbol{a} \cdot \boldsymbol{b} = a^\alpha b^\beta \, \boldsymbol{e}_\alpha \cdot \boldsymbol{e}_\beta = g_{\alpha\beta} a^\alpha b^\beta. \tag{6.2}$$

If the basis is ON, then $g_{\alpha\beta} = \eta_{\alpha\beta}$. Notice that $g_{\alpha\beta}$ is symmetric in that $g_{\alpha\beta} = g_{\beta\alpha}$ (similarly for $T_{\alpha\beta}$, the components of T), while the components $F_{\alpha\beta}$ are antisymmetric in that $F_{\alpha\beta} = -F_{\beta\alpha}$.

If a tensor has type (0, 1), it is called a **covector**. As we know from elementary linear algebra, the set of all covectors forms a vector space (this is called the **dual** of V and represented by V^*), which also has dimension 4. The metric provides a linear, one-to-one correspondence between vectors and covectors where $\boldsymbol{a} \leftrightarrow \alpha$ ($\boldsymbol{a} \in V$, $\alpha \in V^*$) if

$$\alpha(\boldsymbol{b}) = g(\boldsymbol{b}, \boldsymbol{a}) \tag{6.3}$$

for all vectors $\boldsymbol{b}$.

Given a vector basis $(\mathbf{e}_\alpha)$, a convenient choice for a covector basis, (e^α), is one satisfying $e^\alpha(\mathbf{e}_\beta) = \delta^\alpha_\beta$. This is called a **dual basis**. If $\boldsymbol{a} = a^\alpha \mathbf{e}_\alpha$ then

$$e^\beta(\boldsymbol{a}) = e^\beta(a^\alpha \mathbf{e}_\alpha) = a^\alpha e^\beta(\mathbf{e}_\alpha) = a^\alpha \delta^\beta_\alpha = a^\beta.$$

We now come to the full definition of a tensor: A tensor t is a mapping, linear in each of its slots, that takes a sequence of covectors, $\alpha, \ldots, \beta$, and a sequence of vectors, $\boldsymbol{a}, \ldots, \boldsymbol{b}$, and gives a number

$$t(\alpha, \ldots, \beta; \boldsymbol{a}, \ldots, \boldsymbol{b}).$$

If t has N covector slots and M vector slots, we say that it has type (N, M). The reason for this apparently unwarranted complication is that vectors themselves may now be considered to be tensors of type (1, 0). The reason why is that a vector, $\boldsymbol{a}$ say, can take a covector, α say, and give a number $\alpha(\boldsymbol{a})$. The vector $\boldsymbol{a}$ thus defines a tensor of type (1, 0).

The set of all tensors of type (N, M) forms a vector space, which we denote by $\mathbf{R}^N_M$. Note that $V = \mathbf{R}^1_0$ and $V^* = \mathbf{R}^0_1$. For consistency we also write $\mathbf{R} = \mathbf{R}^0_0$. A given tensor t defines a set of related tensors, called its **symmetry family**, obtained by rearranging slots of the same type. For example, if $t \in \mathbf{R}^2_1$, its symmetry family consists of two tensors t and $\hat{t}$ where $\hat{t}(\alpha, \beta; \boldsymbol{a}) = t(\beta, \alpha; \boldsymbol{a})$. In general, if $t \in \mathbf{R}^N_M$, its symmetry family will consist of $N!M!$ tensors. If these tensors are not all distinct, then t is said to have a symmetry.

Given a vector basis $(\mathbf{e}_\alpha)$ and a covector basis (e^α), the components of a tensor t, for example of type (2, 1), are given by

$$t^{\alpha\beta}{}_\gamma = t(e^\alpha, e^\beta; \mathbf{e}_\gamma). \tag{6.4}$$

The ordering of upstairs and downstairs indices plays no role, so we shall write $t^{\alpha\beta}{}_\gamma = t^\alpha{}_\gamma{}^\beta = t_\gamma{}^{\alpha\beta}$ etc.

The **tensor product** st of t with a tensor s of type (0, 1) is of type (2, 2) and defined by

$$st(\alpha, \beta; \boldsymbol{a}, \boldsymbol{b}) = s(\boldsymbol{a})t(\alpha, \beta; \boldsymbol{b}).$$

Similarly ts is defined by

$$ts(\alpha, \beta; \boldsymbol{a}, \boldsymbol{b}) = t(\alpha, \beta; \boldsymbol{a})s(\boldsymbol{b}).$$

Notice that $st \neq ts$. The components of st are $s_\alpha t^{\beta\gamma}{}_\delta$, and the components of ts are $t_\alpha{}^{\beta\gamma} s_\delta$. The definition of a tensor product between any two tensors should now be obvious. Clearly, the tensor product of a tensor of type (N, M) with a tensor of type (N', M') gives a tensor of type $(N+N', M+M')$.

We say that a tensor is **simple** if it can be expressed as a tensor product of vectors and covectors. For example, if there exist vectors $\boldsymbol{v}$ and $\boldsymbol{u}$ and a covector μ such that

$$t = \boldsymbol{v}\boldsymbol{u}\mu,$$

that is, $t(\alpha, \beta; \boldsymbol{a}) = \alpha(\boldsymbol{v})\beta(\boldsymbol{u})\mu(\boldsymbol{a})$, then t is simple. The tensor t admits two **contractions**. The first is with respect to its first covector slot and its vector slot and gives the contracted tensor

$$\mathrm{cr}_1\, t = \mu(\boldsymbol{v})\,\boldsymbol{u},$$

which has type $(1, 0)$; the second is with respect to its second covector slot and its vector slot and gives

$$\mathrm{cr}_2\, t = \mu(\boldsymbol{u})\,\boldsymbol{v},$$

which also has type $(1, 0)$. Even if t is not simple, it can still be expressed as a linear combination of simple tensors, and, by defining the contraction operations to be linear, its contractions can be defined in the obvious way. Life, however, is not quite that simple, because a tensor can be expressed as a linear combination of simple tensors in many different ways; but a little thought shows that its contractions are the same no matter how it is expressed. In general, a contraction of a tensor of type (N, M) is a tensor of type $(N-1, M-1)$.

In terms of simple tensors formed from basis vectors, t has the form

$$t = t^{\alpha\beta}{}_\gamma \boldsymbol{e}_\alpha \boldsymbol{e}_\beta e^\gamma,$$

where $t^{\alpha\beta}{}_\gamma$ are its components given by equation (6.4). This can be seen as follows:

$$t(e^\varepsilon, e^\sigma; \boldsymbol{e}_\lambda) = t^{\alpha\beta}{}_\gamma e^\varepsilon(\boldsymbol{e}_\alpha) e^\sigma(\boldsymbol{e}_\beta) e^\gamma(\boldsymbol{e}_\lambda) = t^{\alpha\beta}{}_\gamma \delta^\varepsilon_\beta \delta^\sigma_\beta \delta^\gamma_\lambda = t^{\varepsilon\sigma}{}_\lambda.$$

We thus have

$$\mathrm{cr}_1\, t = t^{\alpha\beta}{}_\gamma e^\gamma(\boldsymbol{e}_\alpha)\boldsymbol{e}_\beta = t^{\alpha\beta}{}_\gamma \delta^\gamma_\alpha \boldsymbol{e}_\beta = t^{\alpha\beta}{}_\alpha \boldsymbol{e}_\beta,$$

and hence the components of $\mathrm{cr}_1\, t$ are $t^{\alpha\beta}{}_\alpha$. Similarly, the components of $\mathrm{cr}_2\, t$ are $t^{\alpha\beta}{}_\beta$. The **identity tensor** is defined by

$$\delta = \delta^\beta_\alpha \boldsymbol{e}_\beta e^\alpha.$$

Notice that δ has only one contraction, which gives $\mathrm{cr}\,\delta = 4$. A tensor $\hat{T}$ of type $(1, 1)$ allows only one contraction operation, and this gives a number $\mathrm{cr}\hat{T} = \hat{T}^{\alpha}{}_{\alpha}$, where $\hat{T}^{\alpha}{}_{\beta}$ are the components of $\hat{T}$.

The one-to-one correspondence, (6.3), between vectors [tensors of type $(1, 0)$] and covectors [tensors of type $(0, 1)$] also gives a one-to-one correspondence between a tensor $\hat{T}$ of type $(1, 1)$ and a tensor T of type $(0, 2)$ defined by

$$T(\boldsymbol{a}, \boldsymbol{b}) = \hat{T}(\alpha, \boldsymbol{b}) \tag{6.5}$$

where $\alpha \leftrightarrow \boldsymbol{a}$. This idea can be extended in an obvious way to give a one-to-one correspondence between tensors of types (N, M) and $(N-1, M+1)$. In the previous chapter the trace operation, $\mathrm{tr}\,T$, was defined on a tensor T of type $(0, 2)$. This, as we have seen, is related to the contraction operation by $\mathrm{tr}\,T = \mathrm{cr}\hat{T} = \hat{T}^{\alpha}{}_{\alpha}$.

In component form, equation (6.3) is $\alpha_{\beta} = g_{\alpha\beta}a^{\alpha}$, where $\alpha \leftrightarrow \boldsymbol{a}$ and equation (6.5) gives $T_{\alpha\gamma} = g_{\gamma\beta}\hat{T}_{\alpha}{}^{\beta}$. From now on we shall use the same symbol to represent two tensors related in this way and write $a_{\beta} = g_{\alpha\beta}a^{\alpha}$ and $T_{\alpha\gamma} = g_{\gamma\beta}T_{\alpha}{}^{\beta}$. Similarly, given $t^{\alpha\beta}{}_{\gamma}$, we write $t_{\lambda}{}^{\beta}{}_{\gamma} = g_{\lambda\alpha}t^{\alpha\beta}{}_{\gamma}$, $t^{\alpha}{}_{\lambda\gamma} = g_{\lambda\beta}t^{\alpha\beta}{}_{\gamma}$. In other words, we use $g_{\alpha\beta}$ to lower indices. In terms of this notation, equation (6.2) becomes $g(\boldsymbol{a}, \boldsymbol{b}) = a_{\alpha}b^{\alpha}$. Notice that if a vector $\boldsymbol{a}$ has components a^{α}, that is, $\boldsymbol{a} = a^{\alpha}\boldsymbol{e}_{\alpha}$, then equation (6.1) implies

$$\boldsymbol{a}\cdot\boldsymbol{e}_{\beta} = a^{\alpha}\boldsymbol{e}_{\alpha}\cdot\boldsymbol{e}_{\beta} = a^{\alpha}g_{\alpha\beta} = a_{\beta}. \tag{6.6}$$

Given the components a_{α} of a covector, we write the components of the corresponding vector as $a^{\beta} = g^{\alpha\beta}a_{\alpha}$. This defines the symmetric quantity $g^{\alpha\beta}$. (Considered as a matrix, $g^{\alpha\beta}$ is the inverse of $g_{\alpha\beta}$.) Thus

$$a^{\beta} = g^{\alpha\beta}a_{\alpha} = g^{\alpha\beta}g_{\alpha\gamma}a^{\gamma},$$

and hence

$$g^{\alpha\beta}g_{\alpha\gamma} = \delta^{\beta}_{\gamma} = g^{\beta}_{\gamma}.$$

Using $g^{\alpha\beta}$, we can now consistently raise indices and write, for example, $t^{\lambda}{}_{\beta}{}^{\gamma} = g^{\lambda\alpha}t_{\alpha\beta}{}^{\gamma}$ and $t_{\alpha}{}^{\lambda\gamma} = g^{\lambda\beta}t_{\alpha\beta}{}^{\gamma}$.

Let us now see again how the trace and contraction operations are related. In component form we have $T(\boldsymbol{a}, \boldsymbol{b}) = T_{\alpha\beta}a^{\alpha}b^{\beta} = T^{\alpha\beta}a_{\alpha}b_{\beta}$, where $T_{\alpha\beta}$ are the components of some quadratic form T. By equation (6.6) this gives

$$T(\boldsymbol{a}, \boldsymbol{b}) = T^{\alpha\beta}(\boldsymbol{e}_{\alpha}\cdot\boldsymbol{a})(\boldsymbol{e}_{\beta}\cdot\boldsymbol{b})$$

and hence $\mathrm{tr}\,T = T^{\alpha\beta}\boldsymbol{e}_{\alpha}\cdot\boldsymbol{e}_{\beta} = T^{\alpha\beta}g_{\alpha\beta} = T^{\alpha}{}_{\alpha}$.

As a brief illustration of how one can make use of tensors in component form, let us return to our discussion of the Maxwell tensor F. In Section 5.5

we saw that the electric part of the field, relative to an observer with four-velocity v^α, is given by

$$E_\alpha = F_{\alpha\beta} v^\beta. \tag{6.7}$$

Notice that the antisymmetry of $F_{\alpha\beta}$ implies $E_\alpha v^\alpha = 0$. In the same section we also extracted the magnetic part of the field from $F_{\alpha\beta}$, but we did this in an apparently nongeometric way. Let us now rectify this situation by using the available spacetime structure, namely the metric $g_{\alpha\beta}$ and the Levi–Civita tensor $\varepsilon_{\alpha\beta\gamma\delta}$, to extract the magnetic part of the field, B_α, from $F_{\alpha\beta}$ and v^α. There is essentially only one way of doing this, and it is given by

$$B_\alpha = \tfrac{1}{2}\varepsilon_{\alpha\beta\gamma\delta} v^\beta F^{\gamma\delta}. \tag{6.8}$$

Notice that the antisymmetry of $\varepsilon_{\alpha\beta\gamma\delta}$ implies $B_\alpha v^\alpha = 0$.

Definitions (6.7) and (6.8) can be related to the results of Section 5.5 by introducing an ON-basis such that $\boldsymbol{v} = \boldsymbol{e}_0$. In terms of such a basis $v^\alpha = (1, 0, 0, 0)$ and equation (6.7) gives $E_i = F_{i0}$. Similarly, equation (6.8) gives $B_i = \frac{1}{2}\varepsilon_{i0jk}F^{jk}$ which, on using the antisymmetry properties of $\varepsilon_{\alpha\beta\gamma\delta}$, gives the same results as Section 5.5.

The **dual** of $F_{\alpha\beta}$ is defined by

$$^*F_{\alpha\beta} = \tfrac{1}{2}\varepsilon_{\alpha\beta\gamma\delta} F^{\gamma\delta}. \tag{6.9}$$

Using this expression, the electric and magnetic parts of $F_{\alpha\beta}$ have the neat form

$$E_\alpha = F_{\alpha\beta} v^\beta, \qquad B_\alpha = {}^*F_{\alpha\beta} v^\beta. \tag{6.10}$$

If $^*E_\alpha$ and $^*B_\alpha$ are the electric and magnetic parts of $^*F_{\alpha\beta}$, then $^*E_\alpha = -B_\alpha$ and $^*B_\alpha = -E_\alpha$. The dual of the dual of $F_{\alpha\beta}$ thus satisfies

$$^{**}F_{\alpha\beta} = -F_{\alpha\beta}. \tag{6.11}$$

Since the Levi–Cevita tensor $\varepsilon_{\alpha\beta\gamma\delta}$ will play an important role in this book, let us conclude this section by considering a few more of its properties. On using the metric to raise the indices of $\varepsilon_{\alpha\beta\gamma\delta}$, we find $\varepsilon^{0123} = -1$ and also

$$\varepsilon_{\alpha\beta\gamma\delta}\varepsilon^{\alpha\beta\gamma\delta} = -24. \tag{6.12}$$

Consider, now, the expression

$$\varepsilon_{\alpha\beta\gamma\delta}\varepsilon^{\lambda\beta\gamma\delta}.$$

If, for example, $\alpha = 0$ and $\lambda = 1$, then

$$\varepsilon_{0\beta\gamma\delta}\varepsilon^{1\beta\gamma\delta} = 0,$$

because any term in the sum represented by this expression would have a repeated index. We thus have

$$\varepsilon_{\alpha\beta\gamma\delta}\varepsilon^{\lambda\beta\gamma\delta} = X\delta^{\lambda}_{\alpha}$$

for some number X. Equation (6.12) together with $\delta^{\alpha}_{\alpha} = 4$ implies $X = -6$, and thus we obtain the useful equation

$$\varepsilon_{\alpha\beta\gamma\delta}\varepsilon^{\lambda\beta\gamma\delta} = -6\delta^{\lambda}_{\alpha}. \tag{6.13}$$

6.2 The Abstract Index Notation

In the previous section we represented a tensor by a single letter. This has the serious drawback of giving no indication of the tensor's type and involves the introduction of new symbols to indicate its various contractions. For example, a tensor t of type $(2, 1)$ admits two contractions, which we denoted by $\mathrm{cr}_1\, t$ and $\mathrm{cr}_2\, t$, the first contraction being with respect to the first upstairs slot and the second with respect to the second upstairs slot. This sort of notation works well for tensors of low type, but becomes extremely cumbersome for tensors of higher type. As we have seen, this problem can be overcome by introducing a basis and considering the components of the tensor. For example, if $t^{\alpha\beta}{}_{\gamma}$ are the components of t, then we know at once that t has type $(2, 1)$ and that the components of its two contractions are given by $t^{\alpha\beta}{}_{\alpha}$ and $t^{\alpha\beta}{}_{\beta}$. However, this component notation also has a serious drawback. If we do not specify how the basis we use is chosen, the equations we write down will be true tensor equations, having basis-independent meaning. However, in some cases it will be convenient to use a particular basis, for example one adapted to the symmetries of the problem under consideration. If we do this, then the equations we write down for the tensor components may be valid only in that basis. It is important to make a clear distinction between equations that hold between tensors and equations for their components that hold only in a special basis. However, this distinction is blurred by component notation. It is therefore desirable to have a tensor notation that does not involve the introduction of a basis.

The Penrose **abstract index notation** [see, for example, Penrose (1968) and Penrose and Rindler (1986)] is a very useful notational device for tensors that has all the advantages of using components but does not require the introduction or even the existence of a basis. The basic idea is to represent a tensor not by a single letter but by several; the first, the **stem**, representing the symmetry family to which the tensor belongs, and the rest, which are either superscripts or subscripts, representing the vector and covector slots of the tensor. For example, in terms of this notation, a symbol such as $t^{a}{}_{bc}$ represents a tensor of type $(1, 2)$, and $t^{a}{}_{cb}$ represents the same tensor but with the downstairs slots reversed. Also, the contractions of $t^a{}_{cb}$ are represented by $t^a{}_{ab}$ and $t^b{}_{ab}$. It is very important to note that the indices a, b, c etc. are simply labels representing the tensor's slots and do not have numerical values.

We start by choosing a potentially infinite set of symbols, L, which we call **abstract indices**. For typographical convenience, we take L to be the set of all lowercase Latin letters (except for i, j, k, l, m, and n, which we reserve for numerical indices taking values 1, 2, and 3). Consider now expressions of the form $t_a{}^{bc}$, $t_b{}^b{}_a$, $t_{abc}{}^{ef}$, $t_{abc}{}^{bc}$, and so on, together with composite expressions of the form $s_e t^e$, $s_e{}^f t_a{}^{bc}$, $s_{ef}{}^{bc} t_{abc}{}^{fe}$, and so on. We do not allow two abstract indices to lie in the same column, nor do we allow repeated upstairs or repeated downstairs indices. The stem letters, t and s in this case, can be any symbols we care to use. Disregarding repeated indices, if two such expressions have the same set of upstairs indices, U say, and the same set of downstairs indices, D say, then we consider them to belong to the same space. For example, among the above expressions, $t_b{}^b{}_a$, $t_{abc}{}^{bc}$, and $s_{ef}{}^{bc} t_{abc}{}^{fe}$ will belong to the same space. If the numbers of elements in U and D are N and M, we take this space to be a distinct but isomorphic copy of $\mathbf{R}^N_M$. Thus $t_b{}^b{}_a$, $t_{abc}{}^{bc}$ and $s_{ef}{}^{bc} t_{abc}{}^{fe}$ belong to a copy of $\mathbf{R}^0_1$. This space may be thought of as simply $\mathbf{R}^0_1$ with a label (in this case, the single abstract index a) attached. Expressions such as $t_b{}^b{}_a = t_{abc}{}^{bc}$ and $t_b{}^b{}_a + t_{abc}{}^{bc}$ are thus meaningful, while expressions such as $v^a + w^b$ are not, since v^a and w^b belong to different spaces. Every allowed formal expression of the above type will thus be isomorphic to some tensor. If, for example, t^a_b is isomorphic to $t \in \mathbf{R}^1_1$, we write $t \leftrightarrow t^a{}_b$.

The rules of the game are as follows (again, for simplicity, we proceed by example): If $t \leftrightarrow t^a{}_b$ and $s \leftrightarrow s^c{}_{gef}$ then $ts \leftrightarrow t^a{}_b s^c{}_{gef}$. The tensor t admits one contraction, which we denote by $t^a{}_a$ (or $t^e{}_e$ for any abstract index e), and the tensor s admits three contractions, which we denote by $s^e{}_{ebc}$, $s^e{}_{aec}$, and $s^e{}_{abe}$. Also, if $\hat{s}$ is obtained from s by swapping its first two downstairs slots, then $\hat{s} \leftrightarrow s^c{}_{egf}$.

We use g_{ab} ($g \leftrightarrow g_{ab}$, where g is the metric) to lower indices (for example, we have $s_{agef} = g_{ac} s^c{}_{gef}$), and g^{ab}, defined by $g_{ae} g^{be} = \delta^b{}_a$, to raise indices (for example, $s^{cb}{}_{ef} = g^{bg} s^c{}_{gef}$). The symbol $\delta^b{}_a$ represents the identity tensor defined by $v^b = \delta^b{}_a v^a$. Note that $g^b{}_a = g_a{}^b = \delta^b{}_a = \delta_a{}^b$.

In terms of the abstract index notation the basis vectors e^α and $\mathbf{e}_\alpha$ can be written e^α_a and e^a_α, and $e^\alpha(\mathbf{e}_\beta) = \delta^\alpha_\beta$ becomes $e^\alpha_a e^a_\beta = \delta^\alpha_\beta$. We thus have, for example, $v^a = v^\alpha e^a_\alpha$, where $v^\alpha = v^a e^\alpha_a$. We also have $e^a_\alpha e^\alpha_b = \delta^a{}_b$, since $e^a_\alpha e^\alpha_b v^b = e^a_\alpha v^\alpha = v^a$. Using $e^\alpha_a e^a_\beta = \delta^\alpha_\beta$ and $e^a_\alpha e^\alpha_b = \delta^a_b$ (note that since $\delta^a{}_b = \delta_b{}^a$, we can replace this symbol by δ^a_b), we have, for example

$$t_{ab}{}^c = t_{\alpha\beta}{}^\gamma e^\alpha_a e^\beta_b e^c_\gamma$$

and

$$t_{\alpha\beta}{}^\gamma = t_{ab}{}^c e^a_\alpha e^b_\beta e^\gamma_c .$$

All tensor equations therefore look the same in abstract index notation as in component form, except that the abstract indices are replaced with numerical indices. The advantage of the abstract form is, of course, that it does not require the introduction of a basis.

A tensor t_{ab} is symmetric if $t_{ab} = t_{ba}$, and antisymmetric if $t_{ab} = -t_{ba}$. The symmetric part of t_{ab} is given by $t_{(ab)} = (t_{ab} + t_{ba})/2$, and its antisymmetric part by $t_{[ab]} = (t_{ab} - t_{ba})/2$. More generally, if $t \in \mathbf{R}^0_N$, we write

$$t_{(ab\cdots c)} = \frac{1}{N!} \sum_\sigma t_{\sigma(ab\cdots c)},$$

$$t_{[ab\cdots c]} = \frac{1}{N!} \sum_\sigma \operatorname{sign}(\sigma)\, t_{\sigma(ab\cdots c)},$$

where the summation is taken over all $N!$ permutations, $\sigma(ab\cdots c)$, of $ab\cdots c$. For example,

$$t_{(abcd)} = \tfrac{1}{6}(t_{abcd} + t_{bca} + t_{cab} + t_{bac} + t_{cba} + t_{acb})$$

and

$$t_{[abcd]} = \tfrac{1}{6}(t_{abcd} + t_{bca} + t_{cab} - t_{bac} - t_{cba} - t_{acb}).$$

If a particular index, b say, is exempt from the permutations, we write, for example, $t_{(a\cdots\hat{b}\cdots c)}$ or $t_{[a\cdots\hat{b}\cdots c)]}$. Thus, for example, $t_{[a\hat{b}c]} = (t_{abc} - t_{cba})/2$.

In abstract index notation, equations (6.12) and (6.13) give $\varepsilon_{abcd}\varepsilon^{abcd} = -24$ and $\varepsilon_{abcd}\varepsilon^{abch} = -6\delta^h_d$, which form part of the following set of equations:

$$\varepsilon_{abcd}\varepsilon^{abcd} = -24, \tag{6.14}$$

$$\varepsilon_{abcd}\varepsilon^{abch} = -6\delta^h_d, \tag{6.15}$$

$$\varepsilon_{abcd}\varepsilon^{abgh} = -4\delta^{[g}_{[c}\delta^{h]}_{d]}, \tag{6.16}$$

$$\varepsilon_{abcd}\varepsilon^{afgh} = -6\delta^{[f}_{[b}\delta^g_c\delta^{h]}_{d]}, \tag{6.17}$$

$$\varepsilon_{abcd}\varepsilon^{efgh} = -24\delta^{[e}_{[a}\delta^f_b\delta^g_c\delta^{h]}_{d]}. \tag{6.18}$$

The following examples illustrate how the abstract index notation is used and give a number of useful results.

Example 6.1 Given an antisymmetric tensor F_{ab} and a four-velocity vector, v^a, we can construct the following tensors: ${}^*F_{ab}$ (the dual of F_{ab}), $E_a = F_{ab}v^b$, and $B_a = {}^*F_{ab}v^b$, where

$${}^*F_{ab} = \tfrac{1}{2}\varepsilon_{abcd}F^{cd}. \tag{6.19}$$

From equation (6.11) we know that the dual of the dual of a tensor is minus that tensor, that is,

$${}^{**}F_{ab} = -F_{ab}. \tag{6.20}$$

By antisymmetry we have $B_a v^a = E_a v^a = 0$. The vectors B_a and E_a are thus spacelike and contain three degrees of freedom each. Since F_{ab} contains six degrees of freedom, B_a and E_a contain the same information as F_{ab}. We now show how F_{ab} can be constructed from B_a and E_a.

Let $B_{ab} = 2B_{[a}v_{b]}$ and $E_{ab} = 2E_{[a}v_{b]}$. Then

$$^*B_{ab} = \varepsilon_{abcd}B^c v^d \quad \text{and} \quad ^*E_{ab} = \varepsilon_{abcd}E^c v^d.$$

By antisymmetry, $B_{ab}v^b = B_a$, $E_{ab}v^b = E_a$, $^*B_{ab}v^b = 0$, and $^*E_{ab}v^b = 0$. This gives

$$F_{ab} = E_{ab} - {}^*B_{ab}, \tag{6.21}$$

because $E_a = (E_{ab} - {}^*B_{ab})v^b$ and

$$B_a = ({}^*E_{ab} - {}^{**}B_{ab})v^b = ({}^*E_{ab} + B_{ab})v^b.$$

Example 6.2 If F_{ab} is a Maxwell field, then E_a and B_a are the electric and magnetic parts of the field according to an observer, O with four-velocity v^a. Let us see what the electric and magnetic parts are according to an observer O' with four-velocity w^a.

The three-velocity V^a of O' relative to O is given by $\gamma^{-1}w^a = v^a + V^a$ and $v_a V^a = 0$ [see equation (4.4)]. Using the relations of the previous example, we get

$$\begin{aligned}\gamma^{-1}E'_a &= F_{ab}(v^b + V^b) = E_a + F_{ab}V^b \\ &= E_a + (E_{ab} - {}^*B_{ab})V^b = E_a - E_b V^b v_a - \varepsilon_{abcd}B^c v^d V^b.\end{aligned} \tag{6.22}$$

In the slow-motion approximation where $\gamma \approx 1$, this reduces to

$$E'_a = E_a - \varepsilon_{abcd}B^c v^d V^b$$

or, equivalently,

$$\boldsymbol{E}' = \boldsymbol{E} - \boldsymbol{B} \times \boldsymbol{V}.$$

Similarly, doing the same calculation for B_a, we get

$$\boldsymbol{B}' = \boldsymbol{B} + \boldsymbol{E} \times \boldsymbol{V}.$$

EXERCISES

6.1 Prove $\varepsilon_{abcd}\varepsilon^{efgh} = -24\delta^{[e}_{[a}\delta^{f}_{b}\delta^{g}_{c}\delta^{h]}_{d]}$ by showing that each side of this equation has the same components with respect to an ON-basis. By taking successive contractions, prove equations (6.14) to (6.17).

6.2 If $k_{abc} = \varepsilon_{abcd}\, j^d$ and $l_a = \varepsilon_{abcd}k^{bcd}$, how are j_a and l_a related?

6.3 The Maxwell energy–momentum tensor is given by

$$4\pi T_{ab} = -F_{ac}F_{bd}g^{cd} + \tfrac{1}{4}g_{ab}F_{de}F^{de}. \qquad \text{(E6.1)}$$

Show that T_{ab} is symmetric and $T^a_a = 0$.

6.4 Prove the identity

$$g_{ab}F_{cd}F^{cd} = 2(F_{ac}F_b{}^c - {}^*F_{ac}{}^*F_b{}^c).$$

6.5 Use the result of the previous question to prove

$$8\pi T_{ab} = -(F_{ac}F_b{}^c + {}^*F_{ac}{}^*F_b{}^c),$$

and hence show that $T_{ab}v^a v^b > 0$ for any four-velocity vector v^a if $F_{ab} \neq 0$.

6.6 If $F_{[ab}F_{c]d} = 0$, show that there exist vectors U_a and V_a such that $F_{ab} = U_{[a}V_{b]}$.

7

Tensor Fields

A **tensor field** T is a smooth, tensor-valued *function* that assigns a tensor $T(p)$ to each point of M. Note that T denotes the field, while $T(p)$ denotes its value at p. Scalar fields [$T(p)$ is a number] and vector fields [$T(p)$ is a vector] are, of course, special cases of tensor fields. Throughout this chapter, symbols such as v^a and T_{ab} will denote tensor *fields*, and the metric and Levi–Cevita tensors should be thought of as being constant tensor fields. Unless otherwise stated, e^a_α will denote a constant ON basis, and x^α will denote the corresponding coordinate functions with respect to some origin. The components of a tensor field, T_{ab} say, will be denoted by an array, $T_{\alpha\beta}$, of scalar fields.

The **position vector field**, $\mathbf{x}$, with respect to some origin point O, is defined by $\mathbf{x}(P) = OP$, where OP is the displacement vector from O to P. The components x^α of $\mathbf{x}$ are given by $\mathbf{x} = \mathbf{e}_\alpha x^\alpha$ or $x^a = e^a_\alpha x^\alpha$.

The **directional derivative**, $\nabla_v T$, of a tensor field T in direction $\mathbf{v}$ is defined by

$$\nabla_v T(p) = \lim_{\varepsilon \to 0} \frac{T(p + \varepsilon \mathbf{v}(p)) - T(p)}{\varepsilon}. \tag{7.1}$$

Since $\nabla_v T$ is linear in $\mathbf{v}$, it defines a tensor ∇T with an extra slot that accepts $\mathbf{v}$. Thus, for example, $\nabla_c T_{ab}$ is defined by $\nabla_v T_{ab} = v^c \nabla_c T_{ab}$. The components of $\nabla_c T_{ab}$ are given by $\partial_\gamma T_{\alpha\beta}$, where ∂_α is the directional derivative along $\mathbf{e}_\alpha$. In the case of a scalar field, $\nabla f \leftrightarrow \nabla_a f$, defined by $\nabla_v f = v^a \nabla_a f$, is called the **gradient** of f. Sometimes we shall write df instead of ∇f. From equation (7.1) it follows immediately that $\nabla_v \mathbf{x} = \mathbf{v}$. In abstract index form, this gives $v^b \nabla_b x^a = v^a = \delta^a_b v^b$, and hence

$$\nabla_b x^a = \delta^a_b, \qquad \text{or} \quad \nabla_b x_a = g_{ab}. \tag{7.2}$$

In the following examples we use the position vector field to construct a number of interesting tensor fields and use equation (7.2) to find their various derivatives.

Example 7.1 Imagine a bomb that explodes at event O, sending out particles in all possible directions with all possible speeds. The four-velocity of a particle at event P [within $N^+(O)$] will point along $\mathbf{x}(P)$

and will thus form a vector field in $N^+(O)$ given by $v^a = x^a\tau^{-1}$, where $\tau^2 = x^a x_a$ and $\tau > 0$. Using equation (7.2), we have $\nabla_a\tau = v_a$ and thus $v^a\nabla_a\tau = 1$. Furthermore,

$$\nabla_b v_a = (g_{ab} - v^a v_b)\tau^{-1}$$

and hence $\nabla_a v^a = 3\tau^{-1}$. The scalar field $\theta_v = -\nabla_a v^a$ is called the **divergence** of v^a and will play a major role in this book.

Example 7.2 Imagine a point particle with four-velocity v^a whose world line passes through O. Vectors tangent to the future-pointing null rays coming from the world line will form a null vector field, which we call l^a. Let $t = v^a x_a$, $r^a = x^a - tv^a$, $r^2 = -r^a r_a$ $(r \geq 0)$, and $e^a = r^a r^{-1}$. Note that e^a is a unit spacelike vector field orthogonal to v^a. The null vector field l^a is given by $l^a = v^a + e^a$. Using equation (7.2), we have $\nabla_a t = v_a$ and $\nabla_a r = -e_a$. Thus $e^a\nabla_a r = l^a\nabla_a r = 1$. Furthermore,

$$\nabla_b l_a = \nabla_b e_a = (g_{ab} - v_a v_b + e_a e_b)r^{-1},$$

and hence $\nabla_a e^a = \nabla_a l^a = 2r^{-1}$ and $\nabla_{[a}e_{b]} = \nabla_{[a}l_{b]} = 0$.

Example 7.3 The same situation as the previous example, but the particle now carries an electric charge e. The (Coulomb) Maxwell field generated by the particle is given by

$$F_{ab} = 2e\frac{v_{[a}e_{b]}}{r^2}. \tag{7.3}$$

Note that the electric part of the field is given by $E_a = F_{ab}v^b = -ee_a r^{-2}$, while the magnetic part,

$$B_a = \tfrac{1}{2}\varepsilon_{abcd}v^b F^{cd},$$

is zero. Using the relations found in the previous example, we find that

$$\nabla^a F_{ab} = \nabla^{a*}F_{ab} = 0, \tag{7.4}$$

which, as we shall see, are the vacuum Maxwell equations.

Example 7.4 Let n^a be a constant null vector field, and s^a a constant spacelike vector field such that $s^a n_a = 0$. Then the tensor field $F_{ab} = s_{[a}n_{b]}\sin n_c x^c$ satisfies Maxwell's equations, (7.4). This is called a plane-wave solution.

7.1 Congruences and Derivations

The definition of a vector field given above depends crucially on M being an affine space, that is, a space where the notion of equivalent displacements makes sense. When we come to consider curved spacetime, we shall not have a natural affine structure, and so our definition of a vector field, along with the notion of equivalent displacements, will cease to apply. In anticipation of this difficulty, we shall, in this section, consider a number of representations of a vector field that are equally valid in a curved-space setting.

A **congruence** is a space-filling family of curves such that every point of M lies on just one curve. A vector field v^a determines a congruence whose curves are tangent to v^a. We call such curves the **streamlines** of v^a. Furthermore, v^a determines a function, λ say, which acts as a parameter along the congruence, by $v^a \nabla_a \lambda = 1$. We thus get a parametrized congruence. The parameter function λ is not, of course, determined uniquely, but only up to $\lambda \to \lambda + f$ where $v^a \nabla_a f = 0$, that is, f is constant along the congruence. In Example 7.1, the streamlines of v^a are all timelike, future-directed lines from O, and the parameter function is τ; in Example 7.2, the streamlines of l^a are all future-directed null rays from the world line, and the parameter function is r.

Conversely, a parametrized congruence (i.e. a congruence with a parameter function, λ say) determines a vector field v^a, where v^a is tangent to the congruence (this fixes v^a up to a factor) and $v^a \nabla_a \lambda = 1$ (this determines the factor).

Clearly, the representation of a vector field by means of a congruence does not depend on the existence of an affine structure and thus applies in a curved-space setting. For example, the lines of latitude on a sphere, with parameter function φ, may be considered as defining a tangent vector field. This vector field is not, of course, well defined at the north or south pole, but this is because φ is not a well-defined function at those points.

Another means of representing a tensor field is by means of a derivation, defined as follows:

Definition 7.1 A **derivation** v is a mapping that takes a smooth function (scalar field) and gives a new smooth function $v(f)$ such that:

(i) $v(f + g) = v(f) + v(g)$.
(ii) $v(fg) = v(f)\,g + f v(g)$. (This is known as the **Leibniz condition.**)
(iii) If k is a constant function, then $v(k) = 0$.

A vector field $\mathbf{v}$ determines a derivation v by $\nabla_{\mathbf{v}} f = v(f)$. As the following theorem shows, the converse of this statement is also true.

Theorem 7.1 *Given a derivation v, then there exists a unique vector field $\mathbf{v}$ such that $\nabla_{\mathbf{v}} f = v(f)$.*

Proof Let $v^\alpha = v(x^\alpha)$, where x^α are coordinate functions, and let $\mathbf{v} = \mathbf{e}_\alpha v^\alpha$. Thus $\nabla_{\mathbf{v}} f = v^\alpha \partial_\alpha f$. We now show that $v(f) = \nabla_{\mathbf{v}} f$ at any given point O. Without loss of generality, we take x^α such that $x^\alpha(O) = 0$. From a result from calculus, we know that f may be written as

$$f = f(O) + x^{\alpha} H_{\alpha}$$

where $H_{\alpha} = \partial_{\alpha} f$ at O. Using the defining properties of v, we have

$$v(f) = v(x^{\alpha}) H_{\alpha} + x^{\alpha} v(H_{\alpha})$$

and hence $v(f) = v^{\alpha} \partial_{\alpha} f = \nabla_v f$ at point O. □

A derivation thus provides a neat equivalent representation of a vector field, which again applies in a curved-space setting. From now on we shall simply refer to a derivation as a vector field and write $v^a \nabla_a f = v(f)$.

Example 7.5 Consider a unit sphere in ordinary three-space with position vector $\boldsymbol{r}$. A point lies on the sphere if $\boldsymbol{r} \cdot \boldsymbol{r} = 1$, and a vector $\boldsymbol{v}$ is tangent to the sphere if $\boldsymbol{v} \cdot \boldsymbol{r} = 0$. A tangent vector *field* $\boldsymbol{v}$ determines a directional derivative operator ∇_v with the properties of a derivation. Conversely, a slight variation of the above theorem shows that a derivation v determines a unique tangent vector field $\boldsymbol{v}$ such that $\nabla_v f = v(f)$ for all smooth functions on the sphere. If, for example, $v(\varphi) = 0$, where (θ, φ) is a spherical coordinate system, then $\boldsymbol{v}$ points along lines of longitude.

7.2 Lie Derivatives

The **Lie bracket** of two vector fields (derivations), v and w, is defined by

$$[v, w](f) = v(w(f)) - w(v(f)). \tag{7.5}$$

A short calculation shows that $[v, w]$ is also a vector field in that it satisfies the defining properties of a derivation. Since $v(f) = v^a \nabla_a f$, we have

$$[v, w](x^{\alpha}) = v(w^{\alpha}) - w(v^{\alpha}) = v^{\beta} \nabla_{\beta} w^{\alpha} - w^{\beta} \nabla_{\beta} v^{\alpha}. \tag{7.6}$$

The **Lie derivative** (in direction v) of a vector field w is defined by

$$\mathcal{L}_v w = [v, w], \tag{7.7}$$

and the Lie derivative of a scalar field f is defined simply by

$$\mathcal{L}_v f = v(f). \tag{7.8}$$

From these two definitions it is easy to see that

$$\mathcal{L}_v(w + u) = \mathcal{L}_v w + \mathcal{L}_v u, \tag{7.9}$$

$$\mathcal{L}_v(f w) = (\mathcal{L}_v f) w + f \mathcal{L}_v w. \tag{7.10}$$

From equation (7.6) we see that

$$\mathcal{L}_v w^a = v^b \nabla_b w^a - w^b \nabla_b v^a. \tag{7.11}$$

The importance of Lie derivatives lies in the fact that their definition does not depend on the existence of an affine structure. As we shall see, they will prove very useful when we come to consider curved spacetime.

The action of a Lie derivative on a covector u_a is defined by

$$\mathcal{L}_v(u_a w^a) = (\mathcal{L}_v u_a) w^a + u_a \mathcal{L}_v w^a,$$

which, since $u_a w^a$ is a scalar, gives

$$(\mathcal{L}_v u_a) w^a = (v^c \nabla_c u_a + u_c \nabla_a v^c) w^a$$

for all w^a, and hence

$$\mathcal{L}_v u_a = v^c \nabla_c u_a + u_c \nabla_a v^c. \tag{7.12}$$

In a similar way $\mathcal{L}_v$ can be defined to act on any tensor. For example, $\mathcal{L}_v T_{ab}$ is defined by

$$\mathcal{L}_v (T_{ab} a^a b^b) = (\mathcal{L}_v T_{ab}) a^a b^b + T_{ab} (\mathcal{L}_v a^a) b^b + T_{ab} a^a (\mathcal{L}_v b^b),$$

which gives

$$\mathcal{L}_v T_{ab} = v^e \nabla_e T_{ab} + T_{eb} \nabla_a v^e + T_{ae} \nabla_b v^e. \tag{7.13}$$

Similarly we have

$$\mathcal{L}_v T_{abcd} = v^e \nabla_e T_{abcd} + T_{ebcd} \nabla_a v^e + T_{aecd} \nabla_b v^e + T_{abed} \nabla_c v^e + T_{abce} \nabla_d v^e.$$

Let us now consider the action of $\mathcal{L}$ on ε_{abcd}. This gives a completely antisymmetric tensor, and since all such tensors are proportional to ε_{abcd}, we may write

$$\mathcal{L}_v \varepsilon_{abcd} = -\theta_v \varepsilon_{abcd}, \tag{7.14}$$

where θ_v is a geometrical scalar depending only on v^a. This scalar field, which is called the **divergence** of v^a, has many important applications. The following theorem tells us how to calculate it:

Theorem 7.2 $\theta_v = -\nabla_a v^a$.

Proof Since $\nabla_e \varepsilon_{abcd} = 0$, we have

$$-\theta_v \varepsilon_{abcd} = \varepsilon_{ebcd} \nabla_a v^e + \varepsilon_{aecd} \nabla_b v^e + \varepsilon_{abed} \nabla_c v^e + \varepsilon_{abce} \nabla_d v^e.$$

Simply contracting this with ε^{abcd} and using equations (6.14) and (6.15) yields the required result. Alternatively, in component form, we have:

$$-\theta_v \varepsilon_{\alpha\beta\gamma\delta} = \varepsilon_{\varepsilon\beta\gamma\delta}\nabla_\alpha v^\varepsilon + \varepsilon_{\alpha\varepsilon\gamma\delta}\nabla_\beta v^\varepsilon + \varepsilon_{\alpha\beta\varepsilon\delta}\nabla_\gamma v^\varepsilon + \varepsilon_{\alpha\beta\gamma\varepsilon}\nabla_\delta v^\varepsilon.$$

Thus

$$\begin{aligned}-\theta_v = -\theta_v \varepsilon_{0123} &= \varepsilon_{\varepsilon 123}\nabla_0 v^\varepsilon + \varepsilon_{0\varepsilon 23}\nabla_1 v^\varepsilon + \varepsilon_{02\varepsilon 3}\nabla_0 v^\varepsilon + \varepsilon_{012\varepsilon}\nabla_3 v^\varepsilon \\ &= \nabla_0 v^0 + \cdots + \nabla_3 v^3 \\ &= \nabla_\alpha v^\alpha,\end{aligned}$$

which, on reverting to abstract indices, gives

$$\theta_v = -\nabla_a v^a. \qquad \square \quad (7.15)$$

A tensor field t is said to be **Lie-propagated** along v^a if $\mathcal{L}_v t = 0$. Let us investigate the geometrical meaning of this condition.

If a function f is Lie-propagated along v^a, then $\mathcal{L}_v f = v(f) = 0$, which means that f is constant along the congruence of v^a. If a vector field w^a is Lie-propagated along v^a then $\mathcal{L}_v w = [v, w] = 0$. (Note that $\mathcal{L}_v v = [v, v] = 0$. A vector is thus Lie-propagated along itself.) In order to see the geometrical meaning of this condition, we introduce the notion of a **connecting vector field.**

Let (τ, y^1, y^2, y^3) be a (local) coordinate system such that $v(\tau) = 1$ (i.e., τ is a parameter function on the congruence of v^a) and $v(y^i) = 0$ (i.e., the functions y^i are constant along the curves of the congruence). To every curve l of the congruence let us assign a corresponding neighboring curve l'. Given a point P on l, there will exist a corresponding point P' on l' with the same parameter value, that is, $\tau(P) = \tau(P')$ (Fig. 7.1). Furthermore, since the functions y^i are constant along the congruence, $y^i(P) - y^i(P')$ will remain constant as P and his brother, P', slide up the congruence. A vector field $\boldsymbol{w}$ is a connecting vector field if there exists a parameter function τ and a correspondence between curves such that, to first order in ε, $\varepsilon\boldsymbol{w}$ connects points with the same parameter value on corresponding curves. This means that $\nabla_w \tau = 0$ and $\nabla_w y^i$ is constant along the congruence, that is, $\nabla_v \nabla_w y^i = 0$. In terms of an *arbitrary* parameter function satisfying $\nabla_v \tau = 1$, we have $\nabla_v \nabla_w \tau = \nabla_v \nabla_w y^i = 0$, or equivalently, $v(w(\tau)) = v(w(y^i)) = 0$. Since

Figure 7.1. Displacements connecting points on l with points on l' having the same parameter value.

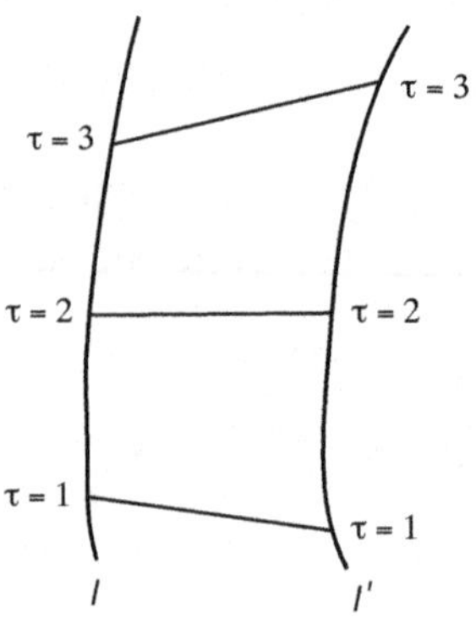

$v(\tau)=1$ and $v(y^i)=0$, we see that w is a connecting vector field if and only if $[v, w]\tau = [v, w]y^i = 0$, which is the case if and only if $[v, w] = 0$. A Lie-propagated vector field is thus a connecting vector field.

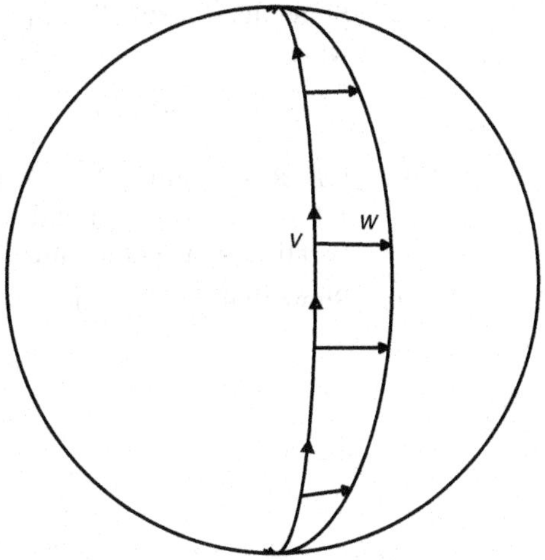

Figure 7.2. A connecting vector w between lines of longitude.

Example 7.6 Let θ and φ be spherical coordinates on a sphere, and let v and w be two vector fields (derivations) such that $v(\varphi)=0$ (v points along lines of longitude), $v(\theta)=1$ (θ is a parameter function), $w(\theta)=0$ (w points along lines of latitude), and $w(\varphi)=\text{const}$ (w connects lines of longitude). Then w is a connecting vector field for the congruence of v, and hence $[v, w] = 0$ (see Fig. 7.2).

A covector α_a that is Lie-propagated along v^a satisfies

$$v(\alpha_a w^a) = \mathcal{L}_v(\alpha_a w^a) = \alpha_a \mathcal{L}_v w^a,$$

which shows that $v(\alpha_a w^a) = 0$ for all connecting vectors w^a. A covector α_a is thus Lie-propagated if and only if $\alpha_a w^a$ is constant along the congruence of v^a for all connecting vectors w^a. Similarly, a tensor T_{ab} is Lie-propagated if and only if $T_{ab}w^a u^a$ is constant along the congruence for all connecting vectors w^a and u^a.

EXERCISES

7.1 Find the magnetic part of a Coulomb field with respect to a *noncomoving* observer.

7.2 A vector potential for a Maxwell field is a vector field A_a such that $F_{ab} = \nabla_{[a}A_{b]}$. Find vector potentials for a Coulomb and a plane-wave Maxwell field.

7.3 For a plane-wave Maxwell field, show that $E^a B_a = 0$.

7.4 Show that the general solution of $\nabla_{(a}\xi_{b)} = 0$ is $\xi_a = M_{ab}x^b + P_a$ where M_{ab} and P_a are constant and M_{ab} is antisymmetric.

7.5 Show that $\mathcal{L}_v w^a = 0$ does not imply $\mathcal{L}_v w_a = 0$.

7.6 Write down the expression for $\mathcal{L}_v T_{ab}{}^c$ in terms of $T_{ab}{}^c$, ∇_a, and v^a.

7.7 Show that $\nabla^{a*}F_{ab} = 0$ implies $\nabla_{[a}F_{bc]} = 0$.

7.8 Show that Maxwell's equations, $\nabla^{a*}F_{ab} = \nabla^a F_{ab} = 0$, imply $\nabla^a T_{ab} = 0$, where

$$4\pi\, T_{ab} = -F_{ac}F_{bd}g^{cd} + \tfrac{1}{4}g_{ab}F_{de}F^{de}.$$

7.9 If $F : \mathbf{R} \to \mathbf{R}$ and $g : M \to \mathbf{R}$ are smooth functions, then $F(g) = F \circ \gamma$ and $F'(g) = F' \circ \gamma$ are smooth functions on M. Prove that any derivation, v, satisfies the **chain rule,** $v(F(g)) = F'(g)v(g)$.

7.10 Show that the Lie bracket of two derivations is a derivation.

8

Field Equations

As we have seen, a (charged) matter distribution on flat spacetime is described by the fields J_a, F_{ab}, and T_{ab}, where J_a determines the charge density, F_{ab} describes the electromagnetic field, and T_{ab} determines the four-momentum density. The background geometry of M determines ∇_a and the constant tensor fields g_{ab} and ε_{abcd}. Nature does not, of course, allow J_a, F_{ab}, and T_{ab} to vary in an arbitrary manner, but imposes constraints in the form of conservation laws and equations of motion, which may be expressed in the form of **field equations**. In this chapter we shall discover the field equations satisfied by J_a, F_{ab}, and T_{ab}.

8.1 Conservation Laws

Consider a congruence whose curves are the particle world lines of a uniform dust distribution with particle-number current j^a. Let us select one particular curve, l_0, and three neighboring curves, l_1, l_2, and l_3, which are joined to l_0 by three space-like connecting vectors a^a, b^a, and c^a (Fig. 8.1).

According to an observer with four-velocity v^a orthogonal to a^a, b^a, and b^a, the volume $V = \varepsilon_{abcd} v^a a^b b^c c^d$ will always contain the same number of particles. Note that v^a is determined uniquely by the condition that it is orthogonal to a^a, b^a, and b^a, but this does not mean that it is Lie-propagated along the congruence. Since

$$\mathcal{L}_j j^a = \mathcal{L}_j a^a = \mathcal{L}_j b^a = \mathcal{L}_j c^a = 0$$

and, by equation (7.14), $\mathcal{L}_j \varepsilon_{abcd} = -\theta_j \varepsilon_{abcd}$, we have

$$\mathcal{L}_j \chi = -\theta_j \chi, \tag{8.1}$$

where $\chi = \varepsilon_{abcd} j^a a^b b^c c^d$. By writing

$$j^a = \rho v^a + A a^a + B b^a + C c^a,$$

we get $j_a v^a = \rho$ and hence

$$\chi = \rho \varepsilon_{abcd} v^a a^b b^c c^d = \rho V = \text{ number of particles in } V. \tag{8.2}$$

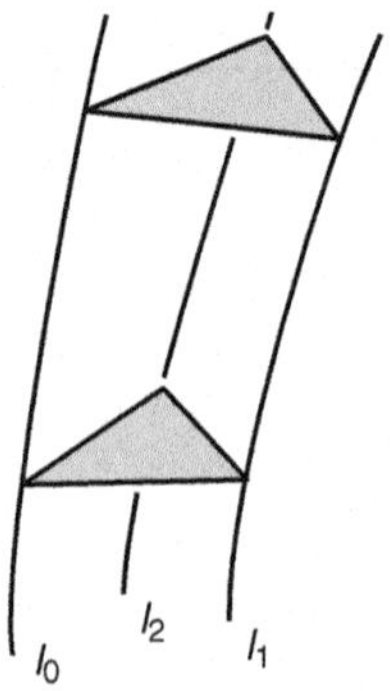

Figure 8.1. As we move up the congruence, the volume spanned by three connecting vectors changes but always captures the same number of particles.

Thus, since the number of particles is constant, we have $\mathcal{L}_j \chi = \mathcal{L}_j(\rho V) = 0$, and hence equations (8.1) and (7.15) give

$$\theta_j = -\nabla_a j^a = 0. \tag{8.3}$$

This is called the particle-number conservation equation. In a similar way, but replacing particle number by charge and four-momentum, we obtain the corresponding conservation equations for charge and energy:

$$\nabla_a J^a = 0 \tag{8.4}$$

and

$$\nabla_a T^{ab} = 0. \tag{8.5}$$

For a perfect fluid,

$$T_{ab} = (\rho + p)u_a u_b - g_{ab} p, \tag{8.6}$$

where u^a is the four-velocity field of comoving particles. In this case, a simple calculation shows that $\nabla_a T^{ab} = 0$ gives the relativistic Euler equation

$$(\rho + p)\dot{u}^a = \nabla^a p - u^a \dot{p} \tag{8.7}$$

and the relativistic continuity equation

$$\dot{\rho} + (\rho + p)\nabla_a u^a = 0. \tag{8.8}$$

Let us now introduce an inertial observer with four-velocity v^a, and write $u^a = \gamma(v^a + V^a)$, where V^a is orthogonal to v^a. The vector field V^a thus gives the 3-velocity of a comoving particle relative to the observer. In the slow-motion approximation, $\gamma \approx 1$, and in three-vector notation we get

$$(\rho + p)\dot{\boldsymbol{V}} = -\boldsymbol{\nabla} p \tag{8.9}$$

and

$$\dot{\rho} + (\rho + p)\,\mathrm{div}\,\boldsymbol{V}, \tag{8.10}$$

which may be recognized as a form of the Euler equation and the matter conservation equation. This justifies our calling p pressure.

8.2 Maxwell's Equations

As we have seen, light or, in other words, electromagnetic radiation plays a central role in spacetime physics. It should come as no surprise, therefore, that the theory of electromagnetism finds its most natural form in such a setting and that, when expressed in tensorial form, its equations assume their simplest and most elegant form. Indeed, as we shall show, Maxwell's equations can be expressed as a single tensorial equation.

A very basic observation is that electric charge is an additive, conserved quantity and comes in two sorts – positive and negative – where like charges repel and unlike charges attract. If an observer releases two identical charged particles from rest, their instantaneous acceleration is given by

$$a = k\frac{e^2}{mr^2} \tag{8.11}$$

where e, m, and r are their charge, rest mass and mutual distance, respectively, and k is some constant. We define our unit of charge by taking this constant to be unity.

Another basic observation is that particles having *magnetic* charge are not found in nature – at least none have, as yet, been observed.

An electromagnetic field makes its presence felt by causing charged particles to accelerate. A particle with charge e, rest mass m, and four-velocity v^a and that instantaneously occupies a point p will have well-defined, instantaneous acceleration at p, which is observed to be proportional to e and inversely proportional to m. The field thus defines a mapping of four-velocity vector fields into vector fields given by

$$v^a \to \frac{m}{e}a^a = E^a(v),$$

where $a^a(p)$ is the acceleration of the particle at p. The vector field $E^a(v)$ is the electric part of the electromagnetic field according to an observer with four-velocity v^a. Since $v^a a_a = 0$, we have $E_a(v)v^a = 0$.

If $E^a(v)$ is produced by a single particle whose world line is the t-axis, then in terms of spherical coordinates, equation (8.11) gives

$$E^a(v) = e\frac{\hat{x}^a}{r^2}, \tag{8.12}$$

where $\hat{x}^a$ is a unit, radial, vector orthogonal to v^a. If this single particle is replaced by a stationary charge distribution with charge density ρ_0, it is a standard exercise in electrostatics to show that

$$\nabla_a E^a(v) = -4\pi\rho_0. \tag{8.13}$$

In standard vector analysis notation, adapted to the 3-plane orthogonal to v^a, this is simply the well-known electrostatic equation

$$\nabla \cdot \mathbf{E} = 4\pi\rho_0.$$

The simplest assumption to make about $E^a(v)$, and one in accord with observations, is that it is *linear* in v^a and hence has the form

$$E_a(v) = F_{ab}v^b \tag{8.14}$$

for some tensor F_{ab} dependent only on the electromagnetic field and hence independent of the observer's motion. Since $E_a(v)v^a = 0$ and $F_{ab}v^a v^b = 0$ for *all* four-velocity vectors v^a, F_{ab} must be antisymmetric. As we have already seen in Chapter 6, the tensor F_{ab} contains a complete description of the electromagnetic field. Let us repeat the argument in a slightly different form.

Being antisymmetric, F_{ab} has six degrees of freedom. Three of these correspond to the electric part of the field, and the remaining three to the magnetic part. In order to extract the magnetic part, we introduce a new antisymmetric tensor

$${}^*F_{ab} = \tfrac{1}{2}\varepsilon_{abcd}\,F^{cd} \tag{8.15}$$

called the *dual* of F_{ab}. It is easily checked that

$${}^{**}F_{ab} = -F_{ab}, \tag{8.16}$$

where ${}^{**}F_{ab}$ is the dual of ${}^*F_{ab}$. In terms of ${}^*F_{ab}$, the magnetic part of the field, according to an observer with four-velocity v^a, is given by

$$B_a(v) = {}^*F_{ab}\,v^b. \tag{8.17}$$

Let us now consider the field equations satisfied by F_{ab}. We begin by using F_{ab} to define two vector fields, J_a^E and J_a^M, according to

$$J_a^E = \nabla^b F_{ab}, \qquad J_a^M = \nabla^{b*} F_{ab}.$$

By antisymmetry, $\nabla^a J_a^E = \nabla^a J_a^M = 0$, and thus J_a^E and J_a^M are conserved currents. Since they are associated with the electric and magnetic parts of the field, it is natural to take J_a^E proportional to the electric current density J_a, and j_a^M proportional to the magnetic current density, which, since magnetic charges do no appear to exist in nature, we take to be zero. The resulting equations are $\nabla^a F_{ab} = kJ_b$ and $\nabla^{a*} F_{ab} = 0$, where k is some constant. In particular, for a stationary distribution with current $J^a = \rho_0 v^a$, this gives $\nabla_a E^a = k\rho_0$, which is consistent with equation (8.13) if we take $k = -4\pi$. We are thus led to the equations

$$\nabla^a F_{ab} = -4\pi\, J_b, \tag{8.18}$$

$$\nabla^{a*} F_{ab} = 0, \tag{8.19}$$

which are precisely **Maxwell's equations** in tensor form.

These equations can be put in the even more concise form

$$\nabla^a W_{ab} = -4\pi J_b \tag{8.20}$$

by introducing a *complex* tensor

$$W_{ab} = F_{ab} + i^* F_{ab}. \tag{8.21}$$

By virtue of equation (8.16), we see that

$$^*W_{ab} = -iW_{ab}, \tag{8.22}$$

$$^*\bar{W}_{ab} = i\bar{W}_{ab}. \tag{8.23}$$

In general, a tensor W_{ab} satisfying (8.22) is called **anti-self-dual**, and a tensor $\bar{W}_{ab}$ satisfying (8.23) is called **self-dual**.

Using the properties of the Levi–Cevita tensor, it can be shown that the tensor T^E_{ab} defined by

$$8\pi T^E_{ab} = -W_{ac}\bar{W}_b{}^c \tag{8.24}$$

is symmetric and real, and satisfies

$$T^{E\,a}_{a} = 0, \tag{8.25}$$

$$\nabla^a T^E_{ab} = -J^a F_{ab}. \tag{8.26}$$

T^E_{ab} is the energy–momentum tensor of the electromagnetic field. It is symmetric, is trace-free (i.e., $T^a_a = 0$),† and satisfies the positivity condition $T^e_{ab}v^a v^b > 0$ for any four-velocity vector v^a. This latter condition can be seen as follows: Making use of the fact that W^{ab} is antisymmetric, we have $W_{ab}v^a v^b = 0$. Furthermore, since

$$g_{ab} = v_a v_b - h_{ab},$$

where h_{ab} is positive definite, we have

$$-8\pi T^E_{ab}v^a v^b = W_{ac}\bar{W}_b{}^c v^a v^b = W^a \bar{W}^b g_{ab} = -W^a \bar{W}^b h_{ab} < 0,$$

where $W_a = W_{ab}v^b$.

In the absence of sources, equation (8.26) implies the conservation law

$$\nabla^a T^E_{ab} = 0,$$

but if $J^a \neq 0$ then T^E_{ab} is not conserved. This is because of the continuous exchange of energy between the source and the field. However, if we take into account the energy–momentum tensor, T^S_{ab}, of the source, then

† This is a common characteristic of energy–momentum tensors associated with massless fields or particles.

the *total* energy–momentum tensor, $T_{ab} = T^E_{ab} + T^S_{ab}$, is conserved. In the case of a uniform flow of charged particles, this can be shown as follows: We have $T^S_{ab} = m_0 v_b j_a$, where j_a is the particle density current, and hence

$$\nabla^a T^S_{ab} = m_0 \nabla^a v_b j_a = m_0 \nabla^a v_b \rho_0 v_a \tag{8.27}$$

$$= m_0 \rho_0 \dot{v}_b = e\rho_0 F_{ba} v^a = F_{ab} J^a. \tag{8.28}$$

Finally, by equation (8.26), we have

$$\nabla^a T_{ab} = \nabla^a T^E_{ab} + \nabla^a T^S_{ab} = 0.$$

8.3 Charge, Mass, and Angular Momentum

As we have seen, the charge and energy density of a relativistic body are determined by its current density, j_a, and its energy–momentum tensor, T_{ab}. In this section we shall use these quantities to obtain expressions for the *total* charge, energy, and angular momentum contained a body that occupies a finite region of space at any instant of time, such as a star. In a spacetime picture, such a body will be described by a world tube, outside of which T_{ab} and j_a vanish (see Fig. 8.2).

We have shown that the charge contained in a small region δV orthogonal to a four-velocity vector t^a is given by $j_a t^a \delta V$. Thus, given a spacelike three-surface V with normal four-velocity vector t^a, the total charge contained in V is given by

$$e = \int_V j_a t^a \, dV. \tag{8.29}$$

Furthermore, using the four-dimensional version of Gauss's theorem, we see that the conservation equation $\nabla_a j^a = 0$ implies that this integral is independent of the choice of V, showing that total charge is conserved.

Figure 8.2. A spacelike three-surface V intersects the world tube of a body in a compact region R.

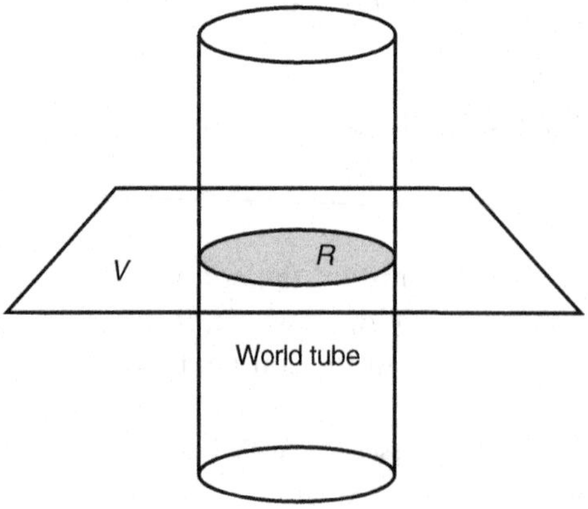

We have also shown that the energy contained in δV with respect to an inertial observer with (constant) four-velocity v^a is given by $T_{ab} v^a t^b \delta V$. Thus, the total energy contained in V is given by

$$E = \int_V T_{ab} v^a t^b \, dV. \tag{8.30}$$

Again, Gauss's theorem together with the conservation equation $\nabla^a T_{ab} = 0$ implies E is conserved in that the integral is independent of the choice of V.

Let us now consider angular momentum. We start with the case of a single particle with constant four-momentum $p^a = mw^a$. Let V now be the orthogonal 3-plane through some point O on the world line of an inertial observer with four-velocity v^a (see Fig. 8.3).

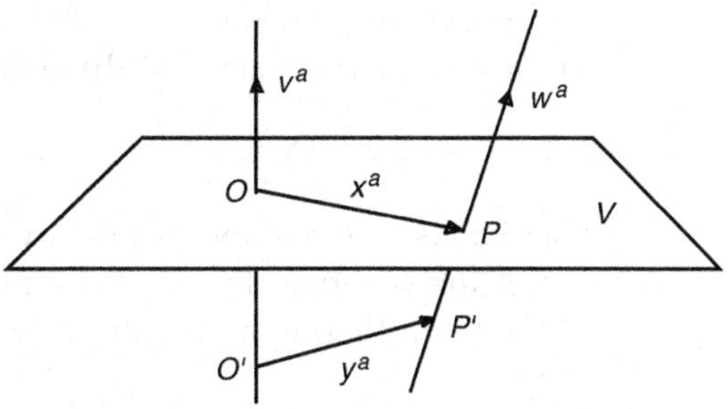

Figure 8.3. V is a 3-plane orthogonal to the world line of an observer with 3-velocity v^a. A particle with four-velocity w^a intersects V at point P.

The angular momentum of the particle with respect to the observer is defined by

$$J_d = -m\varepsilon_{abcd}\, v^a x^b w^c = -\varepsilon_{abcd}\, v^a x^b p^c, \tag{8.31}$$

where x^a connects O to the point P where the world line of the particle intersects V. Given a unit vector e^a orthogonal to v^a, the component of J_a along e^a is given by

$$J = \mathbf{e} \cdot \boldsymbol{J} = -J_d e^d = m\varepsilon_{abcd}\, v^a x^b w^c e^d = \varepsilon_{abcd}\, v^a x^b p^c e^d. \tag{8.32}$$

Note that, by the antisymmetry of ε_{abcd}, we have $J_d v^d = J_d w^d = 0$, and hence J_d is a spacelike vector orthogonal to both v^a and w^a. Furthermore, since any other connecting vector can be written in the form

$$y^a = x^a + \alpha v^a + \beta w^a,$$

J_d remains unchanged if x^a is replaced by y^a.

With respect to the observer, the three-momentum P^a of the particle is given by

$$p^a = E v^a + P^a$$

where $E = p_a v^a$ and $P_a v^a = 0$. Substituting this into equation (8.31), we get

$$J_a = -\varepsilon_{abcd}\, v^a x^b P^c,$$

which, in three-vector notation, gives the usual Newtonian expression for angular momentum,

$$\boldsymbol{J} = \mathbf{x} \times \boldsymbol{P}.$$

The minus sign vanishes because of the $(+---)$ signature of the metric.

In order to give equation (8.32) a neater and more useful form, we introduce an **axisymmetric Killing field**, χ_a, defined by

$$\chi_c = -\varepsilon_{abcd}\, v^a x^b e^d, \tag{8.33}$$

where x^a is a position vector field with origin at any point on the world line of the observer. By the antisymmetry of ε_{abcd}, we see that $\chi_a v^a = \chi_a e^a = 0$. Furthermore, since $\nabla_e x^b = \delta_e^b$, $\nabla_e \chi_c = -\varepsilon_{aecd} v^a e^d$, and hence

$$\nabla_{(e}\chi_{c)} = 0. \tag{8.34}$$

In three-vector notation, equation (8.33) becomes

$$\boldsymbol{\chi} = \mathbf{x} \times \mathbf{e},$$

from which we see that

$$\boldsymbol{\chi} \cdot \boldsymbol{\chi} = -\chi_a \chi^a = r^2 \sin^2\theta.$$

In an axisymmetric situation with axis of symmetry $\mathbf{e}$, $\boldsymbol{\chi}$ will point along the circles of symmetry, and hence any axisymmetric scalar f will satisfy $\boldsymbol{\chi} \cdot \boldsymbol{\nabla} f = 0$.

In terms of χ_a, equation (8.32) becomes

$$J = -m\chi_a w^a = -\chi_a p^a, \tag{8.35}$$

where the right-hand side is understood to be evaluated on the world line of the particle. As we shall see shortly, the same equation also holds in an axisymmetric situation in general relativity and defines the angular momentum of an inertial particle about the axis of symmetry.

In the case of a continuous matter distribution, the four-momentum contained in a small region δV is given by $T_{ab} v^b \delta V$ and, by equation (8.35), this will have angular momentum $-T_{ab}\chi^a v^b \delta V$. The total angular momentum in direction e^a contained in V is thus given by

$$J = -\int_V T_{ab}\chi^a v^b \, dV. \tag{8.36}$$

Unlike the corresponding expressions for total energy and charge, this expression is defined with respect to some chosen 3-plane orthogonal to the world line of the observer. However, using equation (8.34), the conservation equation $\nabla^b T_{ab} = 0$ and the fact that T_{ab} is symmetric, we have

$$\nabla^b (T_{ab}\chi^a) = T_{ab}\nabla^b \chi^a = 0.$$

Thus $T_{ab}\chi^a$ is a conserved current, and by Gauss's theorem we have

$$J = -\int_V T_{ab}\chi^a t^b \, dV \tag{8.37}$$

for *any* choice of three-surface V with normal vector t^a. Angular momentum defined in this way is therefore conserved. It is interesting to note that this conservation law is a consequence of the symmetry of T_{ab}.

In an axisymmetric situation, there will exist a special world line (center-of-mass world line) and a special unit spacelike vector e^a (axis of symmetry) such that the corresponding axisymmetric Killing field χ^a satisfies $\chi^a \nabla_a f = 0$ for all physically defined scalars (energy density, pressure, etc.). In this case, the total angular momentum about the center-of-mass world line will point along e^a, and J will give the total **intrinsic angular momentum**, or **spin**, of the matter distribution.

When we come to the problem of defining energy and intrinsic angular momentum for a stationary, axisymmetric situation in general relativity, we shall be obliged to use $T_{ab} - Tg_{ab}/2$ rather than simply T_{ab}. For a stationary dust distribution, we have $T_{ab} = \rho v_a v_b$ and hence $T = T_a{}^a = \rho$. Thus, $(T_{ab} - Tg_{ab}/2)v^a v^b = \rho/2$ and hence

$$E = 2\int_V (T_{ab} - \tfrac{1}{2}Tg_{ab})v^a v^b \, dV = \int_V \rho \, dV. \tag{8.38}$$

On the other hand, for a axisymmetric distribution, $\chi_a v^a = 0$ implies

$$J = -\int_V (T_{ab} - \tfrac{1}{2}Tg_{ab})\chi^a v^b \, dV = -\int_V T_{ab}\chi^a v^b \, dV. \tag{8.39}$$

The crucial thing here is the factor of 2 in the first equation and the minus sign in the second equation.

In many situations, both in special and general relativity, we have a conserved current j_a, which can be written in the form

$$\nabla^a f_{ab} = j_b \tag{8.40}$$

for some antisymmetric field (2-form) f_{ab}. In such a situation we can use the following form of the three-dimensional Gauss's theorem to express the total charge of j_a as a two-surface integral of f_{ab}.

Theorem 8.1 *If V is a spacelike three-surface with unit normal vector v^a and boundary S, then*

$$\int_V v^b \nabla^a f_{ab} \, dV = \int_S f_{ab} v^a n^b \, dS, \tag{8.41}$$

where f_{ab} is any 2-form and n^a is the unit, outward-pointing (points out of V) normal to S such that $v^a n_a = 0$.

If V is a 3-plane, then this reduces to the usual three-space version of Gauss's theorem,

$$\int_V \boldsymbol{\nabla} \cdot \mathbf{e} \, dV = \int_S \mathbf{e} \cdot \boldsymbol{n} \, dS,$$

where $e^a = f^a{}_b v^b$. The great advantage of using Gauss's theorem in the form given by equation (8.41) is that it carries over unchanged to a curved spacetime setting. This is because it follows from the general form of Stokes's theorem [see, for example, Wald (1984)], which is metric-independent.

Using Gauss's theorem in the case where $\nabla^a f_{ab} = j_b$, we see that the total charge e of j_a is given by

$$e = \int_V j_b v^b \, dV = \int_S f_{ab} v^a n^b \, dS. \tag{8.42}$$

EXERCISES

8.1 Look up estimates for the mass–energy density ρ and the pressure p at the center of the sun. Convert to natural units and show that $\rho \gg p$.

8.2 Given a four-velocity vector v^a, the corresponding charge density ρ_0 and three-current K^a are given by $J^a = \rho_0 + K^a$ where K^a is orthogonal to v^a. Express Maxwell's equations in three-vector form in terms of ρ_0, K^a, E^a, and B^a.

8.3 If $F_{ab} = \nabla_{[a} A_{b]}$, show that A_a can be chosen such that $\nabla_b \nabla^b A_a = 0$ in a region where $J^a = 0$.

8.4 For a perfect fluid, show that $\nabla_a T^{ab} = 0$ implies

$$(\rho + p) u^a \nabla_a u^b = h^{bc} \nabla_c p,$$

where $h_{ab} = g_{ab} - u_a u_b$.

PART THREE

CURVED SPACETIME AND GRAVITY

9

Curved Spacetime

So far we have concentrated on a region of spacetime where gravitational effects may be neglected. Such a region could be the interior of a spaceship hurtling toward the earth over a period of a few seconds, or some vast region of interstellar space. The basic idea is that gravitational tidal effects may be made arbitrarily small by restricting attention to a sufficiently small region of spacetime. This idea is known as the **principle of equivalence**.

We shall now impose no such restriction on the size of our region and consider the geometry of spacetime as a whole, taking into account gravitational tidal effects. This means that we no longer have a physically defined affine structure applicable to the whole of spacetime, and hence no notion of parallel spacetime displacements. We do, however, retain the notion of spacetime points, world lines, null rays, and null cones. Using these physical notions we shall in the next few chapters consider the physical geometry of spacetime in the presence of gravity.

9.1 Spacetime as a Manifold

At its most basic level, spacetime is no more than a set, M, whose points represent the spacetime positions of physical events. A real-valued function f on M assigns a number $f(p)$ to each point p of M. A curve on M may be represented by a one-to-one mapping $c: I \to M$, where I is either an interval (open curve) or a circle (closed curve), which gives a point, $c(t) \in M$ for each $t \in I$. Given a function f and a curve c, we have a function $f_c: I \to \mathbb{R}$, given by $f_c(t) = f(c(t))$.

There exists in nature a special set of curves and a special set of functions that are, in some very basic sense, *smooth*. The world line of an inertial particle should, for example, be taken to be a smooth curve, and scalar fields arising from some naturally evolving process should be taken to be smooth functions. In order to deserve their name, the set of smooth curves and the set of smooth functions must, at the very least, satisfy the following two properties:

- f is a smooth function if and only if f_c is C^∞ for all smooth curves.
- c is a smooth curve if and only if f_c is C^∞ for all smooth functions.

The set of *all* functions together with the *empty* set of curves satisfies these conditions, as does the set of *all* curves together with the set of all constant functions. These two extremes would lead to very boring and rather silly universes. In order to get something in between and that captures the physical notion of smoothness, we need to impose an extra condition that reduces either the size of the set of "smooth" functions or the size of the set of "smooth" curves. The following condition does the trick:

- Given any two distinct points p and q, there exists a smooth curve that joins them and a smooth function such that $f(p) \neq f(q)$.

A set M together with a set of functions, $\mathcal{R}$, and a set of curves, $\mathcal{C}$, satisfying the above properties is called a **smooth space**, and $(\mathcal{R}, \mathcal{C})$ is called a **smooth structure** on M. We shall assume that spacetime is a smooth space and that inertial world lines and null rays are smooth curves.

Example 9.1 The space $\mathbb{R}^2$ is smooth with smooth curves and functions defined in the usual way. This does *not* induce† a smooth structure on $\{(x, y) : y = |x|\}$, because the points $(1, 1)$ and $(-1, 1)$ cannot be joined by a smooth curve. The usual smooth structure on $\mathbb{R}^3$ does, however, induce a smooth structure on the cone $\{(x, y, z) : z = \sqrt{x^2 + y^2}\}$.

Given a point p with parameter value t_p on a smooth curve c, we have a mapping v_p, defined by $v_p(f) = \dot{f}_c(t_p)$, which takes a smooth function f and gives a number $v_p(f)$. The number $v_p(f)$ is, of course, the directional derivative of f along the curve at the point p. We say that the mapping v_p is a **tangent vector** at the point p.

Spacetime is four-dimensional in the sense that is takes four parameters to fix the position of a point. To make this idea precise, we need the notion of an **open set** and the notion of a **chart**.

Definition 9.1 A subset O of M is **open** if the image of any smooth curve through $p \in O$ has a segment in O that contains p (Fig. 9.1). A four-dimensional chart is an open set O together with an ordered set of four smooth functions, $\mathbf{x} = (x^0, x^1, x^2, x^3)$, called a **coordinate system** on O, such that:

(i) $p = q$ if and only if $\mathbf{x}(p) = \mathbf{x}(q)$.
(ii) The image of the mapping $\mathbf{x} : O \to \mathbb{R}^4$ is an open set in $\mathbb{R}^4$.

† Let M be a smooth space and $N \subset M$. If the set of smooth curves that lie in N, and the restriction to N of all smooth functions, form a smooth structure on N, we say that this is **induced** by M.

If every point of a smooth space is contained in a n-dimensional chart, it is said to be a smooth n-dimensional **manifold**.† We assume that spacetime is a smooth four-dimensional manifold.

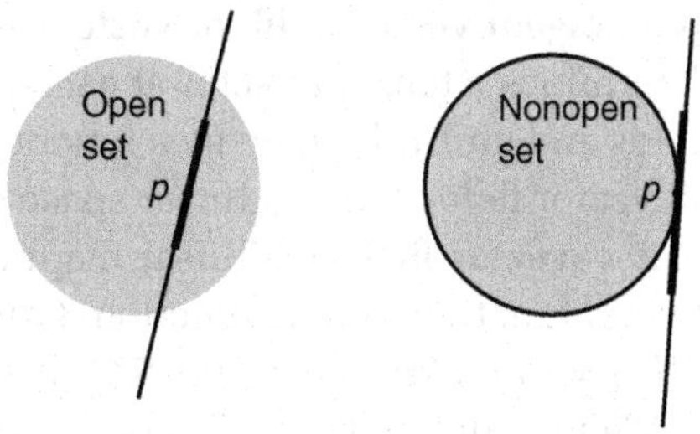

Figure 9.1. An open and a nonopen set.

Example 9.2 A two-sphere, $S^2 = \{(x, y, z) : x^2 + y^2 + x^2 = 1\}$, is a two-dimensional manifold where x, y, and z are smooth functions. The ordered set (x, y, z) satisfies condition (i) but not (ii), and therefore does not qualify as a coordinate system. The ordered set (x, y) does, however, give a coordinate system on the open set given by $z > 0$.

Example 9.3 A cone $\{(x, y, z) : z = \sqrt{x^2 + y^2}\}$ is a smooth space, but not a manifold, because it does not admit a chart containing the vertex. If such a chart existed, z would be a smooth function of x and y.

Since we no longer have the luxury of a natural affine structure, we are forced to define a vector field by means of a smooth congruence, or by means of a derivation. As we shall see, the two notions are equivalent. A congruence is a family of curves such that every point lies on at most one curve of the family. A congruence gives a mapping that takes a smooth function and gives a function $v(f)$, where

$$v(f)(p) = \begin{cases} \dot{f_c}(p) & \text{if } p \text{ lies on a curve,} \\ 0 & \text{otherwise.} \end{cases} \tag{9.1}$$

We say the congruence is smooth if $v(f)$ is smooth for all smooth functions f. From now on we consider only smooth objects and therefore drop the adjective "smooth."

We recall that a derivation is a mapping v that takes a function f and gives a new function $v(f)$ such that:

(i) $v(f+g) = v(f) + v(g)$.
(ii) $v(fg) = (v(f))g + fv(g)$.
(iii) If k is a constant function then $v(k) = 0$.

A congruence thus gives a derivation. Conversely, it can be shown that *all* derivations arise in this way. From now on we shall refer to a derivation

† The definition of a manifold presented here is nonstandard. However, if a topological condition known as paracompactness is imposed, the definition given here is equivalent to the standard one.

as a **tangent vector field**, or vector field for short. Note that a vector field determines a tangent vector at each point of M. Since we can add vector fields and multiply them by a function to get a new vector field, the set of all vector fields forms a linear space over the set of smooth functions.

A **covector field** is a linear mapping α that takes a vector field v and gives a function $\alpha(v)$. A function f gives a covector df (or ∇f) defined by $df(v) = v(f)$. We call df (or ∇f) the **gradient** of f.

We now define tensor fields in the same way as we defined tensors in Chapter 6, the only difference being that our "vector space" is now the linear space of all vector fields over the space of (smooth) *functions* on M. We also introduce the abstract index notation in the same way as before. For example, we denote a vector field v by v^a and a covector field α by α_a. In particular, we denote the covector field df by $\nabla_a f$. Note that $\nabla_a f$ can also be defined by $v(f) = v^a \nabla_a f$ for all vector fields v^a. The differential operator ∇_a, which we call the **gradient** operator, acts only on functions. However, as we shall soon show, it has a natural, unique extension to tensors if M possesses a metric tensor. In this case ∇_a is called a **covariant derivative**.

Let us now restrict attention to some chart O, and use the coordinate functions to construct a basis of vector fields. A function f on M determines a function F of four variables on the image of O in $\mathbb{R}^4$ by $f = F(x^0, x^1, x^2, x^3)$, where

$$F(x^0, x^1, x^2, x^3)(p) = F(x^0(p), x^1(p), x^2(p), x^3(p)).$$

Using this fact, we define a set of derivations (vector fields) ∂_α by

$$\partial_\alpha f = F_\alpha(x^0, x^1, x^2, x^3),$$

where F_α is the partial derivative of F with respect to its αth slot. Thus $\partial_\alpha f(p)$ is the directional derivative of f at p along the αth coordinate curve through p. Note that $\partial_\alpha x^\beta = \delta_\alpha^\beta$ and $\partial_\alpha \partial_\beta f = \partial_\beta \partial_\alpha f$. Given four functions v^0, v^1, v^2, and v^3, $v = v^\alpha \partial_\alpha$ is a vector field such that $v(x^\alpha) = v^\alpha$. In fact, it can be shown that *all* vector fields can be written in this form and hence the fields $e_\alpha = \partial_\alpha$ form a basis of vector fields on O. Thus, on O, a vector field can be expressed as $v = v^\alpha e_\alpha$, or, in abstract index form, $v^a = v^\alpha e_\alpha^a$.

The covector fields $e^\alpha = dx^\alpha$ form a dual basis in that $e^\alpha(e_\beta) = \partial_\beta x^\alpha = \delta_\beta^\alpha$, or, in abstract form, $e_a^\alpha e_\beta^a = \delta_\beta^\alpha$. We can now use the results of Chapter 6 to express any tensor in component form.

Given a tensor, for example t_{ab}, we can define a derivative, $\partial_c t_{ab}$, of t_{ab} by taking $\partial_c t_{ab}$ to be the tensor with components $\partial_\gamma t_{\alpha\beta}$. The differential operator, ∂_a defined in this way, acts on tensors, satisfies the Leibniz rule, is **torsion-free** in that $\partial_a \partial_b f = \partial_b \partial_a f$, and provides an extension of the gradient operator in that $\partial_a f = \nabla_a f$. However, it is not unique, in that it depends on the choice of coordinate functions. If we are dealing with flat spacetime and choose a flat coordinate system, then ∂_a coincides with the operator, ∇_a, defined in Chapter 6, and thus $\partial_c g_{ab} = 0$. If, however, we define ∂_a with respect to curvilinear coordinates, then $\partial_c g_{ab} \neq 0$, because the components $g_{\alpha\beta}$ of the metric will be nonconstant in terms of such coordinate system.

9.2 The Spacetime Metric

There exist three physically defined types of curves on M: timelike curves, which are the world lines of massive particles and which come with a preferred parameter, namely proper time; null geodesics, which are null rays or world lines of massless particles; and spacelike curves, which are everywhere neither timelike nor null. The set of all null geodesics through a point p generates a null cone, $N(p)$; all timelike curves through p fill the inside of $N(p)$, and all spacelike curves fill the outside of $N(p)$. We say that a tangent vector is timelike, null, or spacelike according to whether it is tangent to a timelike curve, a null curve, or a spacelike curve. If, in the timelike case, proper time is the parameter, we say that it is a four-velocity tangent vector.

Using the results of Chapter 2, which still apply locally, we see that there exists a unique metric tensor, g_{ab}, such that $g_{ab}x^a x^b$ is positive, zero, or negative according to whether x^a is timelike, null, or spacelike, and, in the particular case of a four-velocity vector, $g_{ab}v^a v^b = 1$. Such a metric is said to have signature $(1, -1, -1, -1)$.

For flat spacetime with flat space coordinates (t, x, y, z), we have

$$g_{ab} = \nabla_a t \nabla_b t - \nabla_a x \nabla_b x - \nabla_a y \nabla_b y - \nabla_a z \nabla_b z,$$

or, equivalently,

$$g = dt^2 - dx^2 - dy^2 - dz^2.$$

If it is to provide a realistic picture of physical spacetime, our model must consist of a little more than just any manifold together with a metric of signature $(1, -1, -1, -1)$. For example, we would like to avoid a model universe with a boundary or an edge, rather like the "edge of the world" that presumably haunts the imagination of flat-earthists. To avoid such a situation, we demand that (M, g) be **inextendible** in that it may not be taken to be a subspacetime of some larger spacetime. The following example shows that it is not always obvious whether a spacetime is extendible or not.

Example 9.4 Consider a spacetime $(\hat{M}, \hat{g})$ with a global coordinate system $\hat{\mathbf{x}} = (\hat{t}, \hat{x}, \hat{y}, \hat{z})$ such that $\hat{t} > 0$, and where

$$\hat{g} = \hat{t}^{-2} d\hat{t}^2 - d\hat{x}^2 - d\hat{y}^2 - d\hat{z}^2.$$

At first sight it appears that this space is extendible, but in fact that is wrong. If we make the coordinate transformation

$$\hat{t} = e^t, \qquad x = \hat{x}, \qquad y = \hat{y}, \qquad z = \hat{z},$$

we find that

$$\hat{g} = dt^2 - dx^2 - dy^2 - dz^2.$$

The space $(\hat{M}, \hat{g})$ is thus the whole of flat spacetime and is therefore inextendible.

It can be shown that the existence of a metric of signature $(1, -1, -1, -1)$ on a manifold M implies the existence of an equivalence class $\{t^a\}$ of everywhere nonzero vector fields such that $g_{ab}t^a t^b > 0$, that is, t^a is timelike at all point of M. [A proof is contained in Steenrod (1951).] As we saw in the introduction, at any given point of spacetime there is a physical distinction between future- and past-pointing timelike vectors. Let us take two points, p and q, and assume that t^a is future-pointing at p. Since p and q can be connected by a smooth curve, we see that (unless we have a physical discontinuity) t^a must also be future-pointing at q. The equivalence class $\{t^a\}$ thus consists of two disjoint subclasses, $\{t^a\}_+$ and $\{t^a\}_-$, consisting of future- and past-pointing vector fields, respectively. In other words, M has a physically determined **temporal orientation**.

Let us take a unit vector field e_0^a belonging to $\{t^a\}_+$ (i.e., e_0^a is a four-velocity vector field) and, at some given point p, complete it to form an ON-basis $(e_0^a, e_1^a, e_2^a, e_3^a)$. Again, as we saw in the introduction, there is a physical distinction between right- and left-handed ON frames (given that e_0^a is future-pointing). Spacetime thus possesses a physically determined **space orientation** at each point P. Let us take our frame to be right-handed and introduce a completely antisymmetric tensor ε_{abcd} at p such that

$$\varepsilon_{abcd} e_0^a e_1^b e_2^c e_3^d = 1. \tag{9.2}$$

As we saw in Section 4.3, this defines ε_{abcd} uniquely, and furthermore the definition is insensitive to the choice of right-handed ON frame (given that e_0^a is future-pointing). In this way we have a unique tensor, ε_{abcd}, physically defined at each point of M.

9.3 The Covariant Derivative

A **covariant derivative**, or **connection**, ∇_a, is an extension of the gradient operator to tensors that satisfies the Leibniz condition, is torsion-free (i.e. $\nabla_a\nabla_b f = \nabla_b\nabla_a f$), and kills the metric in that $\nabla_c g_{ab} = 0$. By using its affine structure, we were able to construct a covariant derivative (connection) on flat spacetime, but here we are faced with the problem of constructing such an object on curved spacetime where global parallelism is not applicable. We have already come across a differential operator, namely ∂_a, that satisfies all the conditions of a covariant derivative except the last (it does not kill the metric). In the proof of the following theorem, we make use of ∂_a to show that a connection exists and is unique.

Theorem 9.1 *There exists a unique extension ∇_a of the gradient operator that is Leibniz, is torsion-free, and kills the metric.*

Proof Let ∇_a be a connection, and let ∂_a be defined with respect to some *fixed* coordinate system.† Since ∂_a and ∇_a coincide when acting on scalars, we have $(\nabla_a - \partial_a)f = 0$. However, in general, $(\nabla_a - \partial_a)t_b$ will be nonzero. Using the above conditions, we have

$$(\nabla_a - \partial_a)(ft_b + r_b) = f(\nabla_a - \partial_a)t_b + (\nabla_a - \partial_a)r_b,$$

and hence $\nabla_a - \partial_a$ is a linear mapping. We therefore have

$$\nabla_a t_b = \partial_a t_b - \Gamma^c{}_{ab} t_c \tag{9.3}$$

for some tensor $\Gamma^c{}_{ab}$. It is easily checked that torsion-freeness implies $\Gamma^c{}_{ab} = \Gamma^c{}_{ba}$. The freedom in the choice of a connection, when acting on a covector, is therefore given by a symmetric tensor $\Gamma^c{}_{ab}$.

Using this tensor, we can, again by the above conditions, determine the action of ∇_a on a tensor of arbitrary type. For example, we have $\nabla_a(t_c v^c) = \partial_a(t_c v^c)$ and hence

$$(\nabla_a t_c)v^c + t_c\nabla_a v^c = (\partial_a t_c)v^c + t_c\partial_a v^c,$$

which implies

$$-\Gamma^e{}_{ac} t_e v^c + t_e \nabla_a v^e = t_c \partial_a v^c$$

and hence

$$\nabla_a v^e = \partial_a v^e + \Gamma^e{}_{ac} v^c. \tag{9.4}$$

Similarly, we find that

$$\nabla_a(t_b r_c) = \partial_a(t_b r_c) - t_e r_c \Gamma^e{}_{ab} - t_b r_e \Gamma^e{}_{ac}$$

and hence

$$\nabla_a g_{bc} = \partial_a g_{bc} - g_{ec}\Gamma^e{}_{ab} - g_{be}\Gamma^e{}_{ac}. \tag{9.5}$$

† The operator ∂_a is well defined only on a chart. However, by piecing together charts, a global version of ∂_a can be constructed, if M is paracompact.

In general we have

$$\nabla_c t^{a\cdots}{}_{b\cdots} = \partial_c t^{a\cdots}{}_{b\cdots} + \Gamma^a{}_{dc} t^{d\cdots}{}_{b\cdots} + \cdots - \Gamma^d{}_{bc} t^{a\cdots}{}_{d\cdots} - \cdots. \tag{9.6}$$

We now use the last condition, $\nabla_c g_{ab} = 0$, to show that ∇_a exists and is unique.

If g_{ab} is the metric, equation (9.5) becomes

$$\nabla_c g_{ab} = \partial_c g_{ab} - \Gamma_{bca} - \Gamma_{acb}, \tag{9.7}$$

which, on permuting indices, gives

$$\begin{aligned}\partial_c g_{ab} &= \Gamma_{bca} + \Gamma_{acb},\\ \partial_a g_{cb} &= \Gamma_{bac} + \Gamma_{cab},\\ \partial_b g_{ac} &= \Gamma_{cba} + \Gamma_{abc}.\end{aligned}$$

Finally, using the fact that $\Gamma_{a[bc]} = 0$, we have

$$\tfrac{1}{2}(\partial_c g_{ab} + \partial_a g_{cb} - \partial_b g_{ac}) = \Gamma_{bac}. \tag{9.8}$$

There thus exists a *unique* connection ∇_a that kills the metric in that $\nabla_c g_{ab} = 0$. □

The geometrical significance of a metric connection is that it gives a propagation law for vectors (parallel propagation) that preserves lengths and angles. We say that a vector w^a is **parallelly propagated** along the congruence of v^a if $Dw^a = v^e \nabla_e w^a = 0$. If w^a and u^a are parallelly propagated, then

$$v(g(w, u)) = Dg(w, u) + g(Dw, u) + g(w, Du) = 0.$$

Thus $g(w, u)$, and hence lengths and angles, are preserved as we move along the congruence.

In terms of a connection, the Lie bracket $[v, w] \leftrightarrow [v, w]^a$ of the vectors $v \leftrightarrow v^a$ and $w \leftrightarrow w^a$ is given by

$$\begin{aligned}[v, w]f &= [v, w]^a \nabla_a f\\ &= v^e \nabla_e (w^a \nabla_a f) - w^e \nabla_e (v^a \nabla_a f)\\ &= (v^e \nabla_e w^a) \nabla_a f - v^e w^a \nabla_e \nabla_a f - (w^e \nabla_e v^a) \nabla_a f + w^e v^a \nabla_e \nabla_a f\\ &= (v^e \nabla_e w^a - w^e \nabla_e v^a) \nabla_a f,\end{aligned}$$

and hence

$$[v, w]^a = v^e \nabla_e w^a - w^e \nabla_e v^a. \tag{9.9}$$

Since $\mathcal{L}_v w = [v, w]$, we have

$$\mathcal{L}_v w^a = [v, w]^a = v^e \nabla_e w^a - w^e \nabla_e v^a. \tag{9.10}$$

Using the Leibniz condition, we can, exactly as in Chapter 6, extend $\mathcal{L}_v$ to act on tensors of arbitrary type.

As we know, the Levi–Cevita tensor satisfies

$$\varepsilon_{abcd}\varepsilon^{abcd} = -24. \tag{9.11}$$

Thus, by the Leibniz condition, we have $\varepsilon_{abcd}v^f\nabla_f\varepsilon^{abcd} = 0$ which, by antisymmetry, gives $v^f\nabla_f\varepsilon_{abcd} = 0$. Since this equation holds for all v^f, we have

$$\nabla_f\varepsilon_{abcd} = 0. \tag{9.12}$$

Using this together with the results of the previous section, we have

$$\mathcal{L}_v\varepsilon_{abcd} = \nabla_a v^e\varepsilon_{ebcd} + \cdots + \nabla_c v^e\varepsilon_{abce},$$

which, on contracting with ε^{abcd}, gives

$$\mathcal{L}_v\varepsilon_{abcd} = -\theta_v\varepsilon_{abcd}, \tag{9.13}$$

where

$$\nabla_a v^a = -\theta_v. \tag{9.14}$$

The scalar field θ_v is called the **divergence** of v^a. These equations are the same as equations (7.14) and (7.15) of Chapter 6 – which, of course, they should be by the principle of equivalence.

9.4 The Curvature Tensor

Every connection has an associated tensor field, $R_{abc}{}^d$, called its **curvature tensor**, which is defined as follows:

Consider the operator $\nabla_{[a}\nabla_{b]}$. Using the defining properties of ∇_a, we see that the mapping defined by $\nabla_{[a}\nabla_{b]}$ is linear when acting on covectors, that is,

$$\nabla_{[a}\nabla_{b]}(v_c + u_c) = \nabla_{[a}\nabla_{b]}v_c + \nabla_{[a}\nabla_{b]}u_c,$$

and, by torsion-freeness,

$$\nabla_{[a}\nabla_{b]}(fv_c) = f\nabla_{[a}\nabla_{b]}v_c.$$

It thus defines a tensor $R_{abc}{}^d$, the curvature tensor of ∇_a, by

$$2\nabla_{[a}\nabla_{b]}v_c = -R_{abc}{}^h v_h. \tag{9.15}$$

For flat spacetime, there exists a (constant) basis e_a^α such that $\nabla_b e_a^\alpha = 0$. Thus $R_{abc}{}^h e_h^\alpha = 0$ and hence $R_{abc}{}^h = 0$. Conversely, it can be shown that $R_{abc}{}^h = 0$ only implies the existence of a basis satisfying $\nabla_b e_a^\alpha = 0$ on an open neighborhood of any given point. A spacetime with zero curvature need not therefore be identical to flat spacetime described in Chapter 4. Indeed, a spacetime with $R_{abc}{}^h = 0$ can have one of a variety of different topologies and is not restricted to be $\mathbb{R}^4$.

As we shall see in the next chapter, it is precisely the curvature tensor that describes a gravitational field, but for now we shall content ourselves with investigating the mathematical properties of $R_{abc}{}^h$. It should be pointed out that the following results apply to any manifold with a metric of arbitrary (nondegenerate) signature.

If we know the curvature tensor, we know how $\nabla_{[a}\nabla_{b]}$ acts on any tensor. Using the Leibniz condition and torsion-freeness, we have for example,

$$\begin{aligned}
\nabla_{[a}\nabla_{b]}(v_e w^f) &= (\nabla_{[a}\nabla_{b]}v_e)w^f + v_e\nabla_{[a}\nabla_{b]}w^f,\\
0 = \nabla_{[a}\nabla_{b]}(v_e w^e) &= (\nabla_{[a}\nabla_{b]}v_e)w^e + v_e\nabla_{[a}\nabla_{b]}w^e,\\
\nabla_{[a}\nabla_{b]}(v_c w_d) &= (\nabla_{[a}\nabla_{b]}v_c)w_d + v_c\nabla_{[a}\nabla_{b]}w_d,\\
&\vdots
\end{aligned}$$

Thus,

$$2\nabla_{[a}\nabla_{b]}t^h = R_{abc}{}^h t^c, \tag{9.16}$$

$$2\nabla_{[a}\nabla_{b]}t_{cd} = -R_{abc}{}^h t_{hd} - R_{abd}{}^h t_{ch}, \tag{9.17}$$

$$2\nabla_{[a}\nabla_{b]}t^d{}_c = -R_{abc}{}^h t^d{}_h + R_{abh}{}^d t^h{}_c. \tag{9.18}$$

By virtue of its definition, R_{abcd} satisfies the identity $R_{(ab)cd} = 0$. The following theorem shows that it also satisfies a number of other less obvious identities:

Theorem 9.2

$$R_{ab(cd)} = 0, \tag{9.19}$$

$$R_{[abc]d} = 0 \quad \textit{(the first Bianchi identity)}, \tag{9.20}$$

$$\nabla_{[a}R_{bc]de} = 0 \quad \textit{(the second Bianchi identity)}. \tag{9.21}$$

Proof Equation (9.17) gives

$$\partial = -2\nabla_{[a}\nabla_{b]}g_{cd} = R_{abc}{}^h g_{hd} + R_{abd}{}^h g_{ch} = R_{abcd} + R_{abdc},$$

which proves the first identity.

By torsion-freeness,

$$0 = 2\nabla_{[a}\nabla_{[b}\nabla_{c]]}f = -R_{[abc]}{}^d\nabla_d f.$$

This gives the second identity, since covectors of the form $\nabla_a f$ span the space of all covectors, V^*.

We prove the last identity as follows: By equation (9.17) we have

$$-2\nabla_{[a}\nabla_{b]}\nabla_c w_d = R_{abc}{}^e\nabla_e w_d + R_{abd}{}^f\nabla_c w_f.$$

On the other hand, using equation (9.15), we have

$$-2\nabla_a\nabla_{[b}\nabla_{c]}w_d = \nabla_a(R_{bcd}{}^e w_e) = w_e\nabla_a R_{bcd}{}^e + R_{bcd}{}^e\nabla_a w_e.$$

Since

$$\nabla_{[a}\nabla_b\nabla_{c]}w_d = \nabla_{[[a}\nabla_{b]}\nabla_{c]}w_d = \nabla_{[a}\nabla_{[b}\nabla_{c]]}w_d,$$

these two equations imply

$$R_{[abc]}{}^e\nabla_e w_d + R_{[ab\hat{d}}{}^f\nabla_{c]}w_f = w_e\nabla_{[a}R_{bc]d}{}^e + R_{[bc\hat{d}}{}^e\nabla_{a]}w_e.$$

The first term on the left-hand side vanishes by equation (9.20), while the second terms on both sides cancel each other. Thus, we obtain, for all w_e,

$$w_e\nabla_{[a}R_{bc]d}{}^e = 0,$$

which gives the last identity. □

We now define the following derived curvature quantities:

$$R_{ab} = R_{aeb}{}^e \qquad \text{(the **Ricci tensor**)}, \tag{9.22}$$

$$R = R^a{}_a \qquad \text{(the **scalar curvature**)}, \tag{9.23}$$

$$G_{ab} = R_{ab} - \tfrac{1}{2}Rg_{ab} \qquad \text{(the **Einstein tensor**)}. \tag{9.24}$$

We show that R_{ab} is symmetric and that G_{ab} satisfies the identity

$$\nabla^a G_{ab} = 0. \tag{9.25}$$

When written out in full, equations (9.19) to (9.21) give

$$R_{abcd} = -R_{abdc} = -R_{bacd} = R_{cdab}, \tag{9.26}$$

$$R_{abcd} + R_{adbc} + R_{acdb} = 0, \tag{9.27}$$

$$\nabla_a R_{debc} + \nabla_c R_{deab} + \nabla_b R_{deca} = 0. \tag{9.28}$$

Let us take the second equation and raise index d and rename it b. This gives

$$R_{abc}{}^b + R_a{}^b{}_{bc} + R_{ac}{}^b{}_b = 0$$

and hence $R_{ac} - R_{ca} = 0$. The Ricci tensor is therefore symmetric. To prove $\nabla^a G_{ab} = 0$, we take the third equation, raise a and rename it d, and raise c and rename it e. This gives

$$\nabla^d R_{deb}{}^e + \nabla^e R_{de}{}^d{}_b + \nabla_b R_{de}{}^{ed} = 0,$$

which can be written

$$2\nabla^d R_{db} - \nabla_b R = 2(\nabla^b R_{db} - \tfrac{1}{2} g_{db}) = 2\nabla^d G_{db} = 0.$$

Note that, since $g_a^a = 4$, $G = G_a^a = -R$ and hence

$$R_{ab} = G_{ab} - \tfrac{1}{2} G g_{ab}. \tag{9.29}$$

If M has dimension n, the symmetries of the curvature tensor imply the following:

(i) if $n = 1$, $R_{abcd} = 0$;
(ii) if $n = 2$, R_{abcd} has one independent component – essentially R;
(iii) if $n = 3$, R_{abcd} has six independent components – essentially R_{ab};
(iv) if $n = 4$, R_{abcd} has twenty independent components – ten of which are given by R_{ab} and the remaining ten by a tensor called the **Weyl tensor**, C_{abcd}.

For $n = 4$, the **Weyl tensor** is given by

$$\begin{aligned} C_{abcd} = R_{abcd} &+ \tfrac{1}{2}(g_{ad} R_{cb} + g_{bc} R_{da} - g_{ac} R_{db} - g_{bd} R_{ca}) \\ &+ \tfrac{1}{6}(g_{ac} g_{db} - g_{ad} g_{cb}) R. \end{aligned} \tag{9.30}$$

By construction it has the same symmetries as R_{abcd}, but in addition satisfies

$$C_{aed}{}^{e} = 0. \tag{9.31}$$

Thus, $R_{abcd} = C_{abcd}$ if $R_{ab} = 0$.

It is interesting to note that the Weyl tensor first makes an appearance for the spacetime dimension four. Also, as we shall see, it is precisely the Weyl tensor that describes a gravitational field in a vacuum region of spacetime (in such a region, $R_{ab} = 0$). Thus, if we lived in a universe where spacetime had only three dimensions, gravity could not exist in a vacuum region – at least according to general relativity.

Let us see what happens if we replace the metric, g_{ab}, by $\hat{g}_{ab} = \omega g_{ab}$ ($\hat{g}^{ab} = \omega^{-1} g^{ab}$), where ω is a constant. The connection would not change since ∇_a would also kill $\hat{g}_{ab}$, and hence there would be no change in the curvature tensor:

$$\hat{R}_{abc}{}^{d} = R_{abc}{}^{d}. \tag{9.32}$$

Furthermore, contracting on b and d, this gives

$$\hat{R}_{ac} = R_{ac}. \tag{9.33}$$

However,

$$\hat{R} = \hat{g}^{ab} \hat{R}_{ab} = \omega^{-1} g^{ab} R_{ab} = \omega^{-1} R, \tag{9.34}$$

and for the Einstein tensor we get

$$\hat{G}_{ab} = G_{ab}. \tag{9.35}$$

Lowering the upstairs index on the curvature tensor, we get

$$\hat{R}_{abcd} = \hat{g}_{de}\hat{R}_{abc}{}^{e} = \omega g_{de}R_{abc}{}^{e} = \omega R_{abcd}, \tag{9.36}$$

and similarly,

$$\hat{C}_{abcd} = \omega C_{abcd}. \tag{9.37}$$

If we now allow ω to be a nonconstant function, the relationships between the hatted and nonhatted quantities involve derivatives of ω and tend to be rather complicated. However, equation (9.37) remains the same. The Weyl tensor thus has the interesting and useful property of transforming in a very simple way under a rescaling of the metric.

9.5 Constant Curvature

Consider a *three-dimensional* space M with a *positive-definite* metric g_{ab} and a Levi–Cevita tensor ε_{abc} – only three indices, because dim $M = 3$. If the curvature tensor R_{abcd} of M does not determine a preferred direction at any given point, M is said to be **isotropic**. For example, the $t =$const surfaces of a universe satisfying the cosmological principle are spaces of this type.

Writing R_{abcd} in the form R_{AB} where $A = [ab]$ and $B = [cd]$, we see that $R_{AB} = R_{BA}$. The curvature tensor thus provides a *symmetric* mapping, $\lambda_A \to R_A^B\lambda_B$, which takes an antisymmetric tensor, λ_{ab}, and gives a new antisymmetric tensor, $R_{ab}{}^{cd}\lambda_{cd}$. By introducing a basis on the six-dimensional space of antisymmetric tensors of this type, R_{AB} may be expressed as a 6×6 matrix. If the eigenvalues of this mapping are not all equal, there will exist a preferred eigenvector, $\lambda_A = \lambda_{ab}$ say, with maximum eigenvalue. Since this would give a preferred vector, $\lambda_a = \varepsilon_{abc}\lambda^{bc}$, in contradiction to the assumption of isotropy, all eigenvalues must be equal and hence the mapping must be proportional to the identity mapping. We thus have $R^{ab}{}_{cd} = -2K\delta_c^{[a}\delta_d^{b]}$ or, equivalently,

$$R_{abcd} = -K(g_{ac}g_{bd} - g_{ad}g_{bc}). \tag{9.38}$$

Furthermore, the Bianchi identities imply $\nabla_a K = 0$, and therefore K is a constant.

From (9.38) we have

$$\begin{aligned} R_{bd} = R_{ab}{}^{a}{}_{d} &= -K(g_a{}^{a}g_{bd} - g_{ad}g_b{}^{a}) \\ &= -K(3g_{bd} - g_{bd}) \\ &= -2Kg_{bd} \end{aligned}$$

and thus

$$R_{ab} = -2Kg_{ab} \quad \text{and} \quad R = -6K. \tag{9.39}$$

A space whose curvature tensor satisfies (9.38) is said to have **constant curvature.** A three-sphere ($K > 0$), a three-dimensional hyperboloid ($K < 0$), and Euclidean three-space ($K = 0$) are examples of spaces of constant curvature. In fact, it can be shown that any three-dimensional space of constant curvature is, *locally*, one of these three types. This is, however, not true globally. For example, without altering the metric, a box in Euclidean space can be made into a three-dimensional torus by identifying its opposite sides. This space is locally the same as Euclidean space but is, of course, very different globally. Another example is projective three-space P^3, which can be defined by identifying antipodal points of S^3. These two spaces are the same locally but have very different global properties.

Example 9.5 Consider the hyperboloid H given by $x^a x_a = 1$ in flat spacetime. At each point p of H we have a normal four-velocity vector given by $v^a(p) = x^a(p)$. We denote the corresponding normal vector field by v^a. Note that $\nabla_a v_b = g_{ab}$. A vector field t^a is tangent if $t^a v_a = 0$. The tensor field $h_{ab} = g_{ab} - v_a v_b$ provides a projection operator into H in that $h^a_b v^b = 0$ and $t^a = h^a_b t^b$ iff t^a is tangent. Given a tensor field, for example $t^a{}_{bc}$, we define its projection into H by $h^d_a h^b_e h^c_f t^a{}_{bc}$. If a tensor is equal to its projection we say it is tangent. In particular, h_{ab} is tangent. It also gives the induced metric on H in that $h_{ab} t^a s^b = g_{ab} t^a s^b$ where t^a and s^a are tangent. Note that for tensors which are tangent it makes no difference whether we raise or lower indices using h_{ab} or g_{ab}.

The induced connection, D_a, on H is defined by

$$D_a t^{\cdots}_{\cdots} = \text{the projection of } \nabla_a t^{\cdots}_{\cdots},$$

where $t^{\cdots}_{\cdots}$ is tangent. That $D_a h_{bc} = 0$ follows from the fact that it is the projection of $\nabla_a h_{bc}$, which is zero. Using $\nabla_a v_b = g_{ab}$, we have $D_b t_c = \nabla_b t_c + t_b v_c$. Thus,

$$\begin{aligned} D_a D_b t_c &= \text{projection of } \nabla_a \nabla_b t_c + \nabla_a (t_b v_c) \\ &= \text{projection of } \nabla_a \nabla_b t_c + t_b g_{ac}, \\ D_{[a} D_{b]} t_c &= \text{projection of } \nabla_{[a} \nabla_{b]} t_c + t_{[b} g_{a]c} = t_{[b} h_{a]c}, \end{aligned}$$

and hence

$$\begin{aligned} R_{abcd} t^d &= t_a h_{bc} - t_b h_{ac} \\ &= (h_{ad} h_{bc} - h_{ac} h_{bd}) t^d. \end{aligned}$$

We thus have

$$R_{abcd} = -(h_{ac} h_{bd} - h_{ad} h_{bc}),$$

which shows that H is a space of constant curvature $K = 1$. However, h_{ab} is negative-definite. The curvature with respect to the positive-definite metric $-h_{ab}$ is thus $K = -1$.

EXERCISES

9.1 If $\dim M = n$, write down a formula for number of independent components of R_{abcd}.

9.2 Show that $\nabla_{[a}\nabla_{b]}$ is a linear mapping when acting on tensors.

9.3 Show that a unit three-sphere is a space of constant curvature with $K = 1$.

9.4 If $R_{ab} = 0$, show that $\nabla^c C_{abcd} = 0$.

9.5 If ∇_a is the connection with respect to g_{ab}, and $\hat{\nabla}_a$ the connection with respect to the conformally rescaled metric $\hat{g}_{ab} = \Omega^2 g_{ab}$, show that

$$\hat{\nabla}_a t_b = \nabla_a t_b - \gamma_a t_b - \gamma_b t_a + g_{ab}\gamma^e t_e, \tag{E9.1}$$

$$\hat{\nabla}_a t_{bc} = \nabla_a t_{bc} - 2\gamma_a t_{bc} - \gamma_b t_{ac} - \gamma_c t_{ba} + g_{ab}\gamma^e t_{ec} + g_{ac}\gamma^e t_{be}, \tag{E9.2}$$

where $\gamma_a = \nabla_a \Omega / \Omega$.

9.6 Using the results of the previous question, prove:

(a) If t_a is null, then $g^{ca} t_c \hat{\nabla}_a t_b = g^{ca} t_c \nabla_a t_b$.

(b) If t_{bc} is antisymmetric, then $g^{ab}\hat{\nabla}_a t_{bc} = g^{ab}\nabla_a t_{bc}$.

(c)

$$\hat{R}_{ab} = R_{ab} + 2\Omega^{-1}\nabla_a\nabla_b\Omega - 4\Omega^{-2}\nabla_a\Omega\nabla_b\Omega + g_{ab}(\Omega^{-2}\nabla_c\Omega\nabla^c\Omega + \Omega^{-1}\nabla_c\nabla^c\Omega). \tag{E9.3}$$

(d)

$$\hat{C}_{abcd} = \Omega^2 C_{abcd}.$$

10

Curvature and Gravity

We come now to the physical interpretation of the curvature tensor R_{abcd}. As we have seen, R_{abcd} arises from the spacetime metric, which in turn arises from the properties of inertial world lines and null rays. If there are no gravitational tidal effects, then spacetime is flat and hence $R_{abcd} = 0$. Conversely, if $R_{abcd} = 0$, then, apart from the possibility of topologies other than $\mathbb{R}^4$, spacetime is flat and there are no gravitational tidal effects. A nonvanishing curvature tensor is thus an indication and a measure of gravity. In this chapter we shall show that a gravitational field is *completely* described by R_{abcd}. But first we need the concept of a geodesic.

10.1 Geodesics

The world line of an inertial particle has the remarkable property of being uniquely determined by its initial four-velocity vector. If we take proper time to be the parameter along the world line, then its tangent vector at any point will be a four-velocity vector. Similarly, a null ray is uniquely determined by specifying a null vector at some point, O say, and the set of all null rays through O forms a null cone $N(O)$. We recall that it was precisely these properties of inertial world lines and null rays that allowed us to define a unique spacetime metric. Furthermore, the metric led to a unique connection and, finally, to the curvature tensor.

A curve $\lambda \to p(\lambda)$ is said to be a **geodesic** if its parameter λ can be chosen such that the corresponding tangent vector to the curve (satisfying $\lambda^a \nabla_a \lambda = 1$) satisfies the **geodesic equation** $D\lambda^a = 0$, where $D = \lambda^b \nabla_b$. In other words, a geodesic is a curve whose tangent vector is parallelly propagated along itself. λ is said to be an **affine parameter**, and λ^a is said to be an **affine tangent vector**.

If a geodesic is timelike (null or spacelike) at a point, then it is timelike (null or spacelike) at all points, because $D(\lambda^a \lambda_a) = 2\lambda^a D\lambda_a = 0$. Since the geodesic equation $D\lambda^a = 0$ gives a first-order differential equation when expressed in terms of coordinates, a geodesic is uniquely determined by specifying its tangent vector at any point on its world line. If this tangent vector is a four-velocity vector, then the tangent vector at any point will be a four-velocity vector and the affine parameter will be proper time.

We thus have the possibility of two types of preferred motion: inertial and geodesic. However, if these differed, the principle of relativity would be violated. For example, the relative acceleration between an inertial and a geodesic world line, both with the same initial four-velocity vector, would determine a preferred spacelike direction. We thus identify inertial world lines and timelike geodesics, and null rays and null geodesics.

A function u is said to be null if its $n_a = \nabla_a u$, which we take to be nonzero, is null, that is, $n^a n_a = 0$. As we shall see, there is a close connection between null geodesics and null functions. By writing $n^a n_a = 0$ in the form $n^a \nabla_a u = 0$, we see that n^a is tangent to the level surfaces of u and that its streamlines are therefore null curves within these surfaces. The streamlines of n^a are called the **null generators** of u. From the condition $n^a n_a = 0$ we also have

$$n^b \nabla_b n^a = \nabla^b u \nabla_b \nabla_a u = \nabla^b u \nabla_a \nabla_b u = \nabla_a (\nabla^b u \nabla_b u)/2 = 0,$$

and hence the null generators of u are in fact null *geodesics* with affine tangent vector $n_a = \nabla_a u$. The following important theorem shows that any given null geodesic is a null generator of some null function.

Theorem 10.1 *If γ is a null geodesic with affine tangent vector n^a, then there exists (at least locally) a null function u such that $n_a = \nabla_a u$ on γ.*

Proof Take a timelike world line l through a point O on γ, and consider the congruence of all future-directed null geodesics emanating from l. The part of γ to the future of O will thus be a curve of the congruence. Let u be the function whose value at a point P is equal to the value of proper time on l at the point where the null geodesic through P intersects l. The level surfaces of u are thus future null cones with vertices on l (Fig. 10.1). Taking the proper time to be zero at O, we see that the level surface $u = 0$ is the future null cone $N^+(O)$ with vertex O. Our null geodesic γ thus lies in $u = 0$, and hence $n^a \nabla_a u = 0$.

We next show that u is a null function. Let n^a now represent the tangent vector to γ at some point P of $N^+(O)$, and let V be the three-dimensional vector space of tangent vectors to $N^+(O)$ at P, that is, the space of vectors such that $w^a \nabla_a u = 0$ at P. Clearly $n^a \in V$, but V can contain no timelike vectors. This is because a timelike curve through P would enter the *interior* of $N^+(P)$ (i.e., the future null cone with vertex P), which contains no points in common with $N^+(O)$. Let $V^\perp$ be the one-dimensional orthogonal complement of V (i.e., the space of vectors orthogonal to all vectors in V). Since $w^a \nabla_a u = 0$ for all $w^a \in V$, we have $\nabla^a u \in V^\perp$. Furthermore $n^a w_a = 0$ for all $w^a \in V$. If this

Figure 10.1. The function u is constant on a null cone with vertex on the world line.

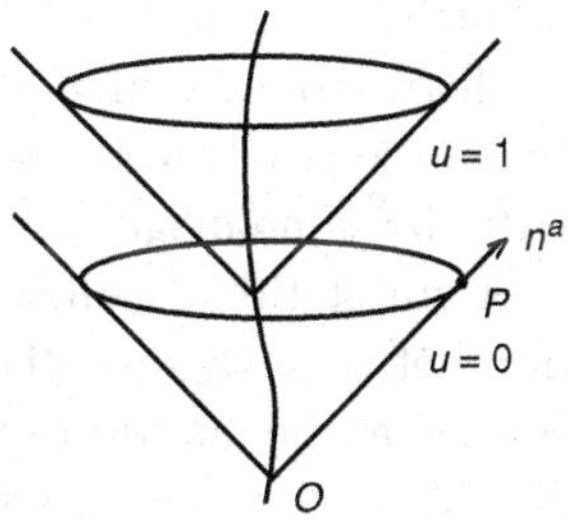

were not the case and there existed a vector $w^a \in V$ such that $n^a w_a \neq 0$, then a linear combination of n^a and w^a could be formed that was timelike. Thus n^a also belongs to $V^\perp$, and hence n^a and $\nabla^a u$ are proportional. Since $n^a \nabla_a u = 0$, this implies $\nabla^a u \nabla_a u = 0$ and hence u is a null function.

Let us now take n^a to be equal to $\nabla^a u$ at P. Since γ and the null generator of u through P have the same tangent vector at P (same initial condition) and both satisfy the geodesic equation, we have the required result that $n^a = \nabla^a u$ on γ. □

What is particularly interesting about this result is that the gradient operation is metric-independent and, moreover, the notion of a null vector is conformally invariant. In other words, if n^a is null with respect to g_{ab}, then it is also null with respect to a conformally related metric $\hat{g}_{ab} = \Omega^2 g_{ab}$. This leads us to the following theorem, which we shall find useful when we come to consider asymptotically flat spacetimes:

Theorem 10.2 *If γ is a null geodesic with respect to g_{ab}, then it is also a null geodesic with respect to $\hat{g}_{ab} = \Omega^2 g_{ab}$. Furthermore, if n^a is an affine tangent vector with respect to g_{ab}, then $\hat{n}^a = \Omega^{-2} n^a$ is an affine tangent vector with respect to $\hat{g}_{ab}$.*

Proof The first part of the theorem follows from the fact any null geodesic is a null generator of some null function and that the notion of a null function is conformally invariant.

To prove the second part we first note that $\hat{\nabla}_a u = \nabla_a u$ (the gradient operation is metric-independent) and $\hat{g}^{ab} = \Omega^{-2} g_{ab}$. From the previous theorem we know that there exists a function u such that $n_a = \nabla_a u$ and $\nabla_a u \nabla^a u = 0$. As we have seen, this implies that $n^a = g^{ab} n_b$ is affine with respect to g_{ab}. If we now let $\hat{n}_a = \hat{\nabla}_a u$, then the same argument implies that $\hat{n}^a = \hat{g}^{ab} \hat{n}_b = \Omega^{-2} g^{ab} n_a = \Omega^{-2} n^a$ is affine with respect to $\hat{g}_{ab}$. □

The **acceleration** of a noninertial observer, N say, is defined to be $D_v v^a$, where $D_v = v^a \nabla_a$. Since $v^a v_a = 1$, we have $D_v v^a v_a = 1$ and thus $D_v v^a$ is spacelike and orthogonal to v^a. At any point P on N's world line, $D_v v^a$ is N's acceleration as measured by an inertial observer, A say, with four-velocity $v^a(P)$ who instantaneously occupies P, that is, both world lines are tangent at P. Being inertial, A has acceleration zero, and therefore the *relative* acceleration of A as measured by N will be given by $a^a = -D_v v^a$. For example, if N is sitting under an apple tree and A is a falling apple (inertial), then a^a points downwards with a magnitude of 32 feet per second squared, and $D_v v^a$, the absolute acceleration of N, points upwards with the same magnitude.

A gravitational field manifests itself by producing a mutual acceleration between two inertial particles, that is, a tidal effect. We shall now show that the *curvature* of spacetime manifests itself in exactly the same way. Consider a congruence of timelike geodesics with tangent vector v^a

satisfying $v_a v^a = 1$ (this, of course, corresponds to a congruence of inertial world lines), and a vector field η^a that is Lie-propagated along the congruence. The vector η^a may thus be viewed as connecting two nearby inertial particles, and $-\eta_a\eta^a$ as the distance between them. Their mutual acceleration may thus reasonably be defined to be $D_v^2\eta^a$. Using equations

$$\mathcal{L}_v\eta^a = v^b\nabla_b\eta^a - \eta^b\nabla_b v^a = 0,$$

$$\nabla_a\nabla_b\eta^c - \nabla_b\nabla_a\eta^c = R_{abe}{}^c\eta^e,$$

we have

$$\begin{aligned}
D_v^2\eta^a &= v^c\nabla_c(v^b\nabla_b\eta^a)\\
&= v^c\nabla_c(\eta^b\nabla_b v^a)\\
&= (v^c\nabla_c\eta^b)\nabla_b v^a + v^c\eta^b\nabla_c\nabla_b v^a\\
&= (\eta^c\nabla_c v^b)\nabla_b v^a + R_{cbd}{}^a\eta^b v^c v^d + v^c\eta^b\nabla_b\nabla_c v^a\\
&= \eta^c\nabla_c(v^b\nabla_b v^a) + R_{cbd}{}^a\eta^b v^c v^d,
\end{aligned}$$

and thus

$$D_v^2\eta^a = \eta^c\nabla_c(v^b\nabla_b v^a) + R_{cbd}{}^a\eta^b v^c v^d. \tag{10.1}$$

Finally, by the geodesic condition $v^b\nabla_b v^a = 0$, we get the important **geodesic deviation equation**

$$D_v^2\eta^a = R_{cbd}{}^a\eta^b v^c v^d. \tag{10.2}$$

This equation expresses the fact that, like gravity, the spacetime curvature causes a mutual acceleration between timelike geodesics. Furthermore, as can easily be seen, the mutual acceleration $D_v^2\eta^a$ is zero for *all* time like geodesic congruences if and only if the curvature tensor vanishes. This is, of course, what we expect from gravity: if no mutual acceleration between any two inertial particles is observed, then the gravitational field is zero. Conversely, if there exist two inertial particles with nonzero mutual acceleration, then they are moving in a nonzero gravitational field. The moral of the story is thus that gravity is curvature and curvature is gravity.

10.2 Einstein's Field Equation

As we have seen, gravitational fields are completely encoded in the curvature tensor R_{abcd} of spacetime, and, furthermore, the energy–mass content of a matter field is completely described by its energy–momentum tensor T_{ab}. Since energy–mass is always accompanied by a gravitational field, we expect some sort of relationship between R_{abcd} and T_{ab}. As a first attempt to find this relationship, we could construct a tensor, T_{abcd} say, out of T_{ab} and g_{ab} that has the same symmetries as R_{abcd} ($T_{abcd} = g_{ad}T_{cb} + g_{bc}T_{da} - g_{ac}T_{db} - g_{bd}T_{ca}$ would do the trick) and simply set $R_{abcd} = kT_{abcd}$. This, however, does not work, as it implies $R_{abcd} = 0$ when $T_{ab} = 0$, and hence no gravitational effects in a vacuum. As a second attempt, we could construct a tensor out of R_{abcd} having the same symmetries as T_{ab}. There are two obvious candidates for this role, namely R_{ab} and G_{ab}. Since $\nabla_a T^{ab} = 0$ while, in general, $\nabla_a R^{ab} \neq 0$, we can reject R_{ab}. We are thus led to consider the equation $G_{ab} = kT_{ab}$. As we shall see, this equation boils down to the Newtonian equation of gravity – under conditions where we expect Newtonian theory to apply – if $k = -8\pi$. The simplest relationship between curvature and matter, consistent with Newtonian gravity, is therefore

$$G_{ab} = -8\pi\, T_{ab}. \tag{10.3}$$

This is called **Einstein's field equation**.

For reasons that we shall consider shortly, Einstein also suggested the following modification to his field equation:

$$G_{ab} = -8\pi\, T_{ab} - \Lambda g_{ab}, \tag{10.4}$$

where Λ is a new universal constant, which has become known as the cosmological term or the **cosmological constant**. This equation is also consistent with Newtonian gravity, but only if Λ is chosen to be very small. It can be shown (Lovelock 1972) that a linear combination of G_{ab} and g_{ab} is the most general two-index symmetric tensor that is divergence-free and can be constructed locally from the metric and its derivatives up to second order. Equation (10.4) therefore gives the most general modification that does not alter the most basic properties of Einstein's field equation.

In many applications it is useful to express Einstein's field equation in terms of R_{ab} rather than $G_{ab} = R_{ab} - \frac{1}{2}Rg_{ab}$. Contracting (10.3), we get $R = -8\pi\, T$ and hence

$$R_{ab} = -8\pi\left(T_{ab} - \tfrac{1}{2}Tg_{ab}\right). \tag{10.5}$$

If we include the cosmological constant, this becomes

$$R_{ab} = -8\pi\left(T_{ab} - \tfrac{1}{2}Tg_{ab}\right) + \Lambda g_{ab}. \tag{10.6}$$

In a sense, Einstein's field equation simply *defines* the energy–momentum tensor corresponding to a given spacetime, just as Newton's law "force = mass × acceleration" simply defines what we mean by force. Any spacetime is thus a solution of Einstein's field equation, but – and this is a big "but"– not all spacetimes have a *physically reasonable* energy–momentum tensor. In order to be physically reasonable, T_{ab} must at least satisfy the weak energy condition

$$T_{ab}w^a w^b \geq 0 \tag{10.7}$$

for any future-pointing vector field w^a, and it is not at all easy to find spacetimes satisfying this condition. If, in addition, we demand that T_{ab} arise from a perfect fluid, we have the constraint

$$T_{ab} = (\rho + p)v_a v_b - p g_{ab}, \tag{10.8}$$

together with the positivity conditions, $\rho \geq 0$ and $\rho \geq -3p$. In this case, Einstein's field equation gives

$$R_{ab} - \tfrac{1}{2} g_{ab} R = -8\pi[(\rho + p)v_a v_b - p g_{ab}],$$

and hence

$$R = -8\pi(3p - \rho), \tag{10.9}$$

$$R_{ab} = -8\pi\left[(\rho + p)v_a v_b + \tfrac{1}{2}(p - \rho)g_{ab}\right], \tag{10.10}$$

$$R_{ab}v^a v^b = -4\pi(3p + \rho). \tag{10.11}$$

Note that for a null fluid ($3p = \rho$), $R = 0$.

Using equation (10.8), it is a straightforward matter to show that the conservation law $\nabla^a T_{ab} = 0$ (or, equivalently, the Bianchi identity $\nabla^a G_{ab} = 0$) is equivalent to the relativistic Euler equation

$$(\rho + p)v^a \nabla_a v^b = (g^{bc} - v^b v^c)\nabla_c p = h^{bc}\nabla_c p \tag{10.12}$$

and the relativistic continuity equation

$$v^a \nabla_a \rho + (\rho + p)\nabla_a v^a = 0. \tag{10.13}$$

Let us now show that, under appropriate circumstances, Einstein's equation leads to Newtonian gravity.

In three-vector notation, Newtonian gravity says this: If $\boldsymbol{a}$ is the acceleration of a freely moving particle in the gravitational field of a (stationary) matter distribution with matter density ρ, then there exists a scalar field φ (the gravitational potential) such that $\boldsymbol{a} = \boldsymbol{\nabla}\varphi$ and $\nabla^2\varphi = 4\pi\rho$. In the language of flat spacetime, this can be reexpressed as follows: Let v^a be the four-velocity of the distribution, and a^a the acceleration of a particle released from rest by a stationary observer (four-velocity v^a). Then there exists a scalar stationary scalar field φ such that $a_a = -\nabla_a\varphi$ and $\nabla^a\nabla_a\varphi = -4\pi\rho$, where $T_{ab} = \rho v_a v_b$. There are minus signs in these expressions because the Euclidean space orthogonal to v^a has a negative-definite

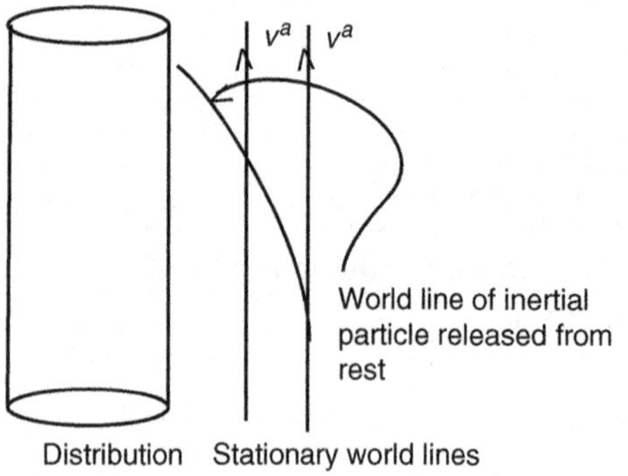

Figure 10.2. An inertial particle accelerates relative to a stationary observer.

metric. For example, in terms of a flat-space coordinate system,

$$\begin{aligned}\nabla^a\nabla_a &= \frac{\partial^2}{\partial t^2} - \frac{\partial^2}{\partial x^2} - \frac{\partial^2}{\partial y^2} - \frac{\partial^2}{\partial z^2}\\ &= \frac{\partial^2}{\partial t^2} - \nabla^2,\end{aligned}$$

and since φ is stationary, $\nabla^a\nabla_a\varphi = -\nabla^2\varphi$.

In the language of curved spacetime, the same physical situation can be described as follows (Fig. 10.2): Let v^a be the four-velocity field tangent to the world lines of observers who consider themselves to be stationary with respect to the distribution, and also stationary with respect to each other in the sense that a connecting vector η^a joining two nearby observers satisfies $D_v^2\eta^a = 0$. Since the particle released from rest by a stationary observer is geodesic (inertial), it has zero acceleration. However, its acceleration *relative* to a stationary observer is given by $a^a = -D_v v^a$, that is, minus the acceleration, $D_v v^a$, of a stationary observer. [Note that $D_v v^a$ must be nonzero; otherwise an (inertial) apple would never have fallen on the head of a *stationary* observer sitting under an apple tree.] If we now take ρ to satisfy $T_{ab} = \rho v_a v_b$ and assume Einstein's equation, the following theorem shows that we get the same equations as in Newtonian gravity.

Theorem 10.3 *Given the curved spacetime situation described above, there exists a stationary scalar field φ such that*

$$a_a = -\nabla_a\varphi \quad \textit{and} \quad \nabla^a\nabla_a\varphi = -4\pi\rho.$$

Proof Since $D_v\eta^a = 0$ for all connecting vectors joining stationary observers, equation (10.1) gives

$$\nabla_e(v^b\nabla_b v^a) = -R_{ced}{}^a v^c v^d,$$

or, equivalently,

$$\nabla_e a_a = R_{ceda} v^c v^d.$$

The symmetries of R_{ceda} give $\nabla_{[e}a_{a]} = 0$, and hence there exists a scalar field φ such that $a_a = -\nabla_a\varphi$. Thus

$$\nabla_e a^a = -\nabla_e\nabla^a\varphi = R_{ced}{}^a v^c v^d,$$

and hence

$$\nabla_a \nabla^a \varphi = R_{cad}{}^a v^c v^d \tag{10.14}$$
$$= R_{cd} v^c v^d \tag{10.15}$$
$$= -8\pi \left(T_{cd} v^c v^d - \tfrac{1}{2} T \right) \tag{10.16}$$
$$= -4\pi\rho, \tag{10.17}$$

where we have used Einstein's equation in the form (10.5) and $T_{ab} = \rho v_a v_b$. □

10.3 Gravity as an Attractive Force

Physically, gravity manifests itself as an *attractive* force in the sense that two nearby inertial particles, initially at relative rest, tend to move toward each other. They may, for instance, be test particles moving in some external gravitational field (e.g., two freely falling particles in the earth's gravitational field), or moving under the influence of their mutual gravitational fields. In this section, we see what Einstein's equation has to say about this phenomenon.

Consider a cloud of inertial dust particles. This will give a timelike, geodesic congruence with a four-velocity tangent vector field v^a and with the proper time t as a parameter function. For simplicity, let us assume that v^a is normal to the $t =$ const surfaces in that $v_a = \nabla_a t$. Thus $\nabla_a v_b = \nabla_b v_a = \nabla_a \nabla_b t$. By writing the metric in the form

$$g_{ab} = v_a v_b + h_{ab},$$

we see that $h_{ab} v^a = 0$ and $g_{ab} a^a b^b = h_{ab} a^a b^b$ if $v_a a^a = v_a b^a = 0$. The tensor h_{ab} is thus the induced 3-metric on the $t =$ const surfaces and has the negative-definite signature $(-1, -1, -1)$. Note that $h_a{}^a = g_a{}^a - v_a v^a = 3$.

The volume spanned by three connecting vectors, a^a, b^a, and c^a orthogonal to v^a is given by

$$V = \varepsilon_{abcd} v^a a^b b^c c^d.$$

Since $\mathcal{L}_v \varepsilon_{abcd} = -\theta \varepsilon_{abcd}$, where $\theta = -\nabla_a v^a$ [see equations (9.13) and (9.14)], and

$$\mathcal{L}_v v^a = \mathcal{L}_v a^a = \mathcal{L}_v b^a = \mathcal{L}_v c^a = 0,$$

we have

$$\mathcal{L}_v V = v(V) = -\theta V. \tag{10.18}$$

The divergence, $\theta = -\nabla_a v^a$, thus determines how a small Lie-propagated region changes volume as we move along the congruence.

A related quantity, called the **shear** of the congruence, is the symmetric tensor σ_{ab} defined by

$$\nabla_a v_b + \tfrac{1}{3}\theta h_{ab} = \sigma_{ab}. \qquad (10.19)$$

This tensor determines the shearing motion of the region as we move along the congruence. If $\sigma_{ab} = 0$, then a small spherical region will remain spherical, but if $\sigma_{ab} \neq 0$, it will be stretched in directions corresponding to the eigenvalues of σ_{ab}.

Since $\nabla_a v^a = -\theta$ and $h_a{}^a = 3$, we have $\sigma_a{}^a = 0$, that is, σ_{ab} is trace-free. Furthermore, since $v^a \nabla_a v^b = 0$ and $v^a h_{ab} = 0$, we have $\sigma_{ab} v^a = 0$ and hence

$$\sigma^{ab}\sigma_{ab} = \sigma^{ab}\sigma^{cd} g_{ca} g_{db} = \sigma^{ab}\sigma^{cd} h_{ca} h_{db} \geq 0. \qquad (10.20)$$

Since h_{ab} is negative-definite, the equality will apply only when $\sigma_{ab} = 0$.

Using the definition of the curvature tensor together with the geodesic condition $D_v v^a = 0$ and the fact that $\nabla_a v_b = \nabla_b v_a$, we have

$$\begin{aligned}
v^c \nabla_c \nabla_b v_a &= v^c \nabla_b \nabla_c v_a - R_{cba}{}^d v^c v_d \\
&= \nabla_b (v^c \nabla_c v_a) - (\nabla_b v^c)(\nabla_c v_a) - R_{cba}{}^d v^c v_d \\
&= -(\nabla_b v^c)(\nabla_c v_a) - R_{cba}{}^d v^c v_d.
\end{aligned}$$

By contracting on the indices b and a this gives

$$v^c \nabla_c \nabla^a v_a = -(\nabla^a v^c)(\nabla_c v_a) - R_c{}^a{}_a{}^d v^c v_d, \qquad (10.21)$$

and, on using (10.19), we get the celebrated **Raychaudhuri equation**

$$v(\theta) = \tfrac{1}{3}\theta^2 + \sigma_{ab}\sigma^{ab} - R_{ab} v^a v^b. \qquad (10.22)$$

Raychaudhuri's equation has many important applications, but for now let us use it to show that gravity, as expressed by Einstein's equation, is an attractive rather than a repulsive force. This is due to the fact the $\sigma_{ab}\sigma^{ab}$ is positive and, by the strong energy condition, $-R_{ab} v^a v^b$ is also positive. Thus $v(\theta) \geq 0$.

Imagine, now, a *stationary* dust cloud in the pitch blackness of intergalactic space, well away from any gravitational field – even that caused by electromagnetic radiation. Since there will be nothing to cause any small volume of dust to expand, contract, or shear, θ and σ_{ab} will be zero and, by the Raychaudhuri equation, will continue to be zero while $R_{abcd} = 0$. If, however, the dust cloud is suddenly illuminated by a passing radio wave, T_{ab} and hence R_{ab} will become nonzero, and, by Raychaudhuri's equation, θ will start to *increase* and become positive. The gravitational influence of the radio wave will also, in general, give the dust particles a shearing motion, and, by equation (10.22), this will only serve to enhance the increase of θ. Any small volume of dust will thus start to *contract* according to the equation $v(V) = -\theta V$, and continue to contract until the dust particles collide along some form of caustic. The pure gravitational influence of a passing star will also have the same effect, except that the contraction will now be caused by C_{abcd} rather than R_{ab}. ($R_{ab} = 0$ in a vacuum region.) This will tend to give the congruence a shearing motion, and the term $\sigma_{ab}\sigma^{ab}$ in Raychaudhuri's equation will cause the congruence to contract.

It is interesting to note the effect the cosmological constant has on this process. By substituting (10.6) into the Raychaudhuri equation we get

$$v(\theta) = \tfrac{1}{3}\theta^2 + \sigma_{ab}\sigma^{ab} + 8\pi\left(T_{ab}v^a v^b - \tfrac{1}{2}T\right) - \Lambda, \tag{10.23}$$

from which we see that Λ has a retarding effect on the contraction and can, in this sense, be interpreted as giving a *repulsive* force.

EXERCISES

10.1 Using the results of Exercise 9.6(a), prove Theorem 10.2.

10.2 Using the results of Exercise 9.6(b), show that the vacuum Maxwell equations,

$$\nabla^a F_{ab} = \nabla^{a*} F_{ab} = 0,$$

are conformally invariant in that

$$\hat{\nabla}^a \hat{F}_{ab} = \hat{\nabla}^{a*} \hat{F}_{ab} = 0,$$

where $\hat{F}_{ab} = F_{ab}$.

10.3 Using the results of Exercise 9.4, show that, in a vacuum region,

$$\nabla^a C_{abcd} = \nabla^{a*} C_{abcd} = 0,$$

where

$${}^*C_{abcd} = \tfrac{1}{2}\varepsilon_{abef}C^{ef}{}_{cd}.$$

10.4 If $\sigma_{ab} = 0$, show that the Raychaudhuri equation reduces to

$$\dot{\theta} = \tfrac{1}{3}\theta^2 + 4\pi(3p + \rho)$$

for a perfect fluid. If $V = a^3$, show that

$$-3\frac{\ddot{a}}{a} = 4\pi(3p + \rho)$$

$[\dot{f} = v(f)]$.

10.5 Given $R_{ab}v^a v^b \le 0$ and that θ takes a *positive* value θ_0 at some point on a geodesic of the congruence (i.e., the congruence is converging at this point), show that θ goes to ∞ along the geodesic within proper time $3/|\theta_0|$.

11

Null Congruences

In the previous section we showed that gravity, as described by the curvature tensor, exerts a converging influence on a uniform flow of inertial particles, given that energy is locally positive. This is in accord with everyday experience, in that gravity is always observed to be an attractive force. Indeed, as we have seen, this effect provides the necessary link between general relativity and Newtonian gravity. We now turn our attention to the effect gravity has on a uniform flow of *massless* particles, that is, a geodesic null congruence. We shall show that, even in this situation, gravity still has a converging influence. This is important, since it highlights the difference between general relativity and Newtonian gravitational theory, where light rays are unaffected by gravity. Indeed, the converging influence of gravity on light rays has led to some of the more remarkable predictions of general relativity, such as black holes and spacetime singularities.

Throughout this chapter we take l^a to be an affine tangent vector to a geodesic null congruence and take r to be an affine parameter such that $Dr = 1$, where $D = l^a \nabla_a$.

11.1 Surface-Forming Null Congruences

A null congruence is **surface-forming** if its tangent vector l^a has the form $l_a = \nabla_a u$ for some null function u. An important example of such a congruence is that formed by future-directed null rays emanating from some timelike world line. Here the level surfaces of u will be future null cones with vertices on the world line. For flat spacetime, Example 7.2 shows that r may be chosen such that its level surfaces in the null cones are two-spheres and that the divergence function ρ, which we define by

$$\rho = -\tfrac{1}{2}\nabla_a l^a. \tag{11.1}$$

(note the $\frac{1}{2}$), is given by

$$\rho = -\frac{1}{r}. \tag{11.2}$$

In the corresponding curved-space situation, the level surfaces of r will be distorted two-spheres and ρ will have the form

$$\rho = -\frac{1}{r} + O(r) \tag{11.3}$$

for some suitable choice of r.

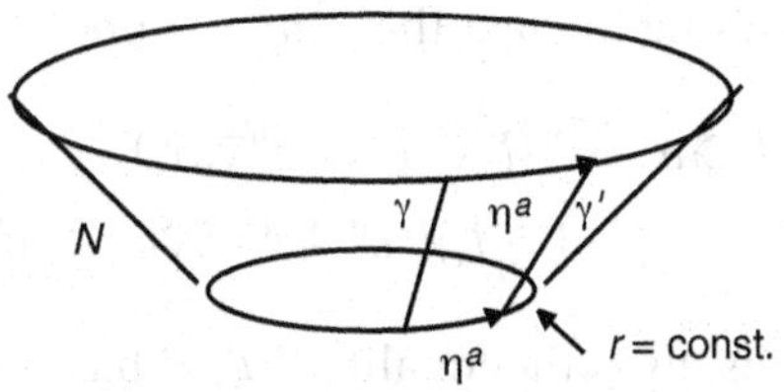

Figure 11.1. The connecting vector η^a joins points in the same level surface N of u with the same r-value.

Returning to the general situation, let λ^a and η^a be two spacelike connecting vectors such that

$$\lambda^a \nabla_a u = \lambda^a \nabla_a r = \eta^a \nabla_a u = \eta^a \nabla_a r = 0,$$

that is, λ^a and η^a join points in the same $u =$ const surface having the same r-value (Fig. 11.1).

At any given point, we define the area spanned by λ^a and η^a to be

$$A = \varepsilon_{abcd} v^a s^b \lambda^c \eta^d,$$

where v^a is a four-velocity vector and s^a is an orthogonal unit spacelike vector, both orthogonal the two-plane spanned by λ^a and η^a. Thus, A is the area spanned by λ^a and η^a according to an observer with four-velocity v^a. There is, of course, considerable freedom in the choice of v^a, but using the antisymmetry of ε_{abcd} it can easily be seen that A is independent of this freedom and depends only on λ^a and η^a.

The following important theorem tells us how A varies as we move up the congruence.

Theorem 11.1 *If A is the area spanned by two Lie-propagated connecting vectors that are tangent to a level surface of u, then*

$$DA = -2\rho A. \tag{11.4}$$

Proof By introducing a null vector n^a such that $l^a n_a = 1$, a convenient choice of v^a and s^a is

$$v^a = \frac{1}{\sqrt{2}}(l^a + n^a) \quad \text{and} \quad s^a = \frac{1}{\sqrt{2}}(l^a - n^a).$$

This gives

$$A = \varepsilon_{abcd} v^a s^b \eta^c \lambda^d = \varepsilon_{abcd} n^a l^b \eta^c \lambda^d. \tag{11.5}$$

By equations (9.13), (9.14), and (11.1) we have $\mathcal{L}_l \varepsilon_{abcd} = -2\rho\varepsilon_{abcd}$. Using this together with the fact that $\mathcal{L}_l l^a = \mathcal{L}_l \lambda^a = \mathcal{L}_l \eta^a = 0$, we get

$$\mathcal{L}_l A = DA = -2\rho A + \varepsilon_{abcd} \mathcal{L}_l n^a l^b \eta^c \lambda^d.$$

It now only remains to show that $\varepsilon_{abcd} \mathcal{L}_l n^a l^b \eta^c \lambda^d = 0$.

Using the fact that $l_a n^a = 1$ and $\nabla_a l_b = \nabla_b l_a$, we have

$$\begin{aligned} l_a \mathcal{L}_l n^a &= l_a (l^b \nabla_b n^a - n^b \nabla_b l^a) \\ &= (\nabla_b l_a) l^b n^a - l^a n^b \nabla_b l_a = 0. \end{aligned}$$

Thus, by orthogonality, $l_a \mathcal{L}_l n^a$ has the form

$$\mathcal{L}_l n^a = a\, l^a + b\eta^a + c\lambda^a,$$

and hence by the antisymmetry of ε_{abcd} we have $\varepsilon_{abcd}\mathcal{L}_l n^a l^b \eta^c \lambda^d = 0$. □

In order to obtain a Raychaudhuri-type equation for a null congruence it is convenient at this stage to complete l^a to a **null tetrad** $(n^a, l^a, m^a, \bar{m}^a)$ (note that m^a is complex) such that $l^a n_a = -m^a \bar{m}_a = 1$ with all other contractions zero. By virtue of these orthogonality relations we have

$$g^{ab} = l^a n^b + n^a l^b - m^a \bar{m}^b - \bar{m}^a m^b.$$

To see this we simply note that the orthogonality relations imply that g^{ab} satisfies $l^a = g^{ab} l_b$, $n^a = g^{ab} n_b$, and $m^a = g^{ab} m_b$. Since $l^a \nabla_a l_b = 0$ and $\nabla_a l_b = \nabla_b l_a$, we have

$$\begin{aligned} \rho &= -\tfrac{1}{2} g^{ab} \nabla_a l_b = -\tfrac{1}{2}(l^a n^b + n^a l^b - m^a \bar{m}^b - \bar{m}^a m^b) \nabla_a l_b \\ &= \bar{m}^a m^b \nabla_a l_b, \end{aligned}$$

which gives the following alternative definition of ρ:

$$\rho = \bar{m}^a m^b \nabla_a l_b. \tag{11.6}$$

Another important function associated with a null congruence is

$$\sigma = m^a m^b \nabla_a l_b. \tag{11.7}$$

This function is called the **shear**, and, like σ_{ab} in the timelike case, gives a measure of the shearing motion of the congruence. When $\nabla_a l_b$ is expanded in terms of a null-tetrad basis, the above definitions of ρ and σ imply

$$\nabla_a l_b = \rho(m_a \bar{m}_b + \bar{m}_a m_b) + \bar{\sigma}\, m_a m_b + \sigma\, \bar{m}_a \bar{m}_b + X l_a l_b \tag{11.8}$$

where X is some irrelevant function. Substituting this into 10.21 (with v^a replaced by l^a), we get

$$D\rho = \rho^2 + \sigma\bar{\sigma} - \tfrac{1}{2} R_{ab} l^a l^b. \tag{11.9}$$

This is called the **Raychaudhuri–Newman–Penrose equation** and (as its name might indicate) has many important applications in general relativity.

A corresponding equation for σ can also be obtained by contracting

$$(\nabla_a \nabla_b - \nabla_b \nabla_a) l_c = -R_{abcd} l^d$$

with $l^a m^b m^c$. This gives

$$D\sigma = 2\rho\sigma + R_{abcd} l^a m^b l^c m^d,$$

and hence, using equation (9.30), we get

$$D\sigma = 2\rho\sigma + C_{abcd} l^a m^b l^c m^d. \tag{11.10}$$

Both the strong and weak energy conditions imply $R_{ab} l^a l^b \leq 0$ and hence $D\rho \geq 0$, which shows that gravity has a converging influence on null particles. If $\rho = \sigma = 0$ initially, then in the absence of a gravitational field they will remain zero. If, however, some form of gravitational influence appears, $-R_{ab} l^a l^b$ will, in general, become positive, and, in addition, the congruence will in general develop a shearing motion. The Raychaudhuri–Newman–Penrose equation thus tells us that ρ will become positive and will inexorably increase. The equation $DA = -2\rho A$ then implies that the area between any three massless particles will inexorably *decrease* until some form of caustic is reached where the area becomes zero.

11.2 Twisting Null Congruences

Let us now briefly consider a null geodesic congruence that is not surface-forming, that is, $\nabla_a l_b \neq \nabla_b l_a$. In this case we say that the congruence is **twisting**. If we now define ρ by equation (11.6) rather than by (11.1) (these two definitions differ if the congruence is not surface-forming) we still get the Raychaudhuri-type equation

$$D\rho = \rho^2 + \sigma\bar{\sigma} - \tfrac{1}{2} R_{ab} l^a l^b,$$

but ρ is now complex, reflecting the fact that the congruence is not surface-forming. However, the corresponding equation for σ is now given by

$$D\sigma = (\rho + \bar{\rho})\sigma + C_{abcd} l^a m^b l^c m^d. \tag{11.11}$$

The congruence is completely determined by its tangent vector l^a, but ρ and σ depend also on the completion of l^a to form a null tetrad. In order to get some handle on the geometrical significance of ρ and σ it is therefore necessary to see how they depend on this completion. Given l^a, a null tetrad is determined up to transformations of the form

$$m^a \to m^a + \bar{A} l^a, \qquad n^a \to n^a + \bar{A} m^a + A\bar{m}^a + A\bar{A} l^a$$

and

$$n^a \to n^a, \qquad m^a \to e^{i\lambda} m^a.$$

Using the fact that $l^a \nabla_a l^b = 0$, it is easy to see that ρ and σ remain unchanged under the first transformation. Under the second transformation, ρ again remains unchanged but σ transforms like $\sigma \to e^{2i\lambda}\sigma$. The quantities ρ and $\sigma\bar{\sigma}$ thus have geometrical significance in that they depend only on l^a and not on its completion to a null tetrad. In particular, if a congruence is shear-free in one completion, then it is shear-free in all completions. Shear-freeness is thus a well-defined geometrical notion applicable to null geodesic congruences.

In Theorem 10.2 we proved that if l^a is null and affine with respect to g_{ab}, then $\hat{l}^a = \Omega^{-2} l^a$ is null and affine with respect to the conformally related metric $\hat{g}_{ab} = \Omega^2 g_{ab}$. In other words, the notion of a geodesic null congruence is conformally invariant. In the following theorem we show the notion of shear is also conformally invariant.

Theorem 11.2 *If σ is the shear of l^a with respect to g_{ab}, then the shear $\hat{\sigma}$ of $\hat{l}^a = \Omega^{-2} l^a$ with respect to $\hat{g}_{ab} = \Omega^2 g_{ab}$ is given by $\hat{\sigma} = \Omega^{-2}\sigma$. In particular, shear-freeness is conformally invariant.*

Proof Here we make essential use of the fact that a Lie derivative is independent of the choice of metric. In particular,

$$\mathcal{L}_m l^a = m^b \nabla_b l^a - l^b \nabla_b m^a = m^b \hat{\nabla}_b l^a - l^b \hat{\nabla}_b m^a.$$

Using this equation together with $\nabla_a(m^b m_b) = 2m^b \nabla_a m_b$, we have

$$m_a \mathcal{L}_m l^a = m_a(m^b \nabla_b l^a - l^b \nabla_b m^a) = m_a m^b \nabla_b l^a = \sigma$$

and, similarly, $\hat{m}_a \mathcal{L}_{\hat{m}} \hat{l}^a = \hat{\sigma}$.

A suitable choice of $\hat{m}^a$ is given by $\hat{m}^a = \Omega^{-1} m^a$, as this gives $\hat{g}_{ab} \hat{m}^a \bar{\hat{m}}^b = -1$. Note that $\hat{m}_a = \Omega m_a$.

Substitution into $\hat{\sigma} = \hat{m}_a \mathcal{L}_{\hat{m}} \hat{l}^a$ now gives

$$\begin{aligned}
\hat{\sigma} &= \Omega m_a \mathcal{L}_{\hat{m}} (\Omega^{-2} l^a) = \Omega^{-1} m_a \mathcal{L}_{\hat{m}} l^a \\
&= \Omega^{-1} m_a (\hat{m}^b \nabla_b l^a - l^b \nabla_b \hat{m}^a) \\
&= \Omega^{-1} m_a [\Omega^{-1} m^b \nabla_b l^a - l^b (\Omega^{-1} m^a)] \\
&= \Omega^{-2} m_a m^b \nabla_b l^a = \Omega^{-2} \sigma.
\end{aligned}$$

□

Let us now consider the transformation behavior of ρ under a conformal rescaling. In the surface-forming case (ρ real) this can easily be obtained from the fact that, on any given geodesic γ, $\rho = -DA/2A$, where A is the area spanned by two connecting vectors joining γ to two nearby geodesics, and $D = l^a \nabla_a$. Under a conformal rescaling, the area spanned by the same two connecting vectors will be given by $\hat{A} = \Omega^2 A$, and, as we have already seen, $\hat{l}^a = \Omega^{-2}$. Thus

$$\hat{\rho} = -\frac{\hat{D}\hat{A}}{2\hat{A}} = -\frac{DA}{2A} - \frac{D\Omega}{\Omega^3} = \rho - \frac{D\Omega}{\Omega^3}.$$

The key point here is that while ρ is not conformally invariant itself (because of the $D\Omega\,\Omega^{-3}$ term), it possesses a conformally invariant *potential*, namely A, such that $\hat{A} = \Omega^2 A$. If we look back at the proof of $\rho = -DA/2A$, we see that it also works formally in the non-surface-forming case but where A is now a *complex* function. Thus, even in the non-surface-forming case, we have

$$\hat{\rho} = \rho - \frac{D\Omega}{\Omega^3}. \tag{11.12}$$

The importance of geodesic, shear-free, null congruences in general relativity is that they enable us to impose a well-defined geometrical condition on a spacetime. Restricting attention to the vacuum case ($R_{ab} = 0$), we say that a spacetime is **algebraically special** if it admits a geodesic, shear-free, null congruence, or, in other words, a null vector field l^a such that $l^a \nabla_a l^b = 0$ and $\sigma = 0$. Unfortunately, unlike other geometrical conditions one can impose on a spacetime, such as asymptotic flatness and stationarity, algebraic specialness has no *direct* physical motivation. It does, however, impose a very rigid and mathematical interesting structure, and this has led to the discovery of several explicit and physically interesting vacuum spacetimes. Indeed, almost all explicit vacuum spacetimes of any physical interest are algebraically special.

The rigidity imposed by algebraic specialness can, in a small way, be appreciated by considering the Raychaudhuri equation, which reduces to simply $D\rho = \rho^2$. By introducing an affine parameter r such that $Dr = 1$, the nontrivial solutions of this equation have the form $\rho = -1/(r + A)$, where A is a *complex* function such that $DA = 0$ (i.e., A is constant along the congruence). The affine parameter condition, $Dr = 1$, defines r up to $r \to r + B$ where B is a *real* function such that $DB = 0$. There thus exists a unique affine parameter and a unique real function Σ such that

$$\rho = -\frac{1}{r + i\Sigma} \tag{11.13}$$

where $D\Sigma = 0$. The function Σ is sometimes called the **twist** of the congruence and, as we shall see, has a close connection with the spin or intrinsic angular momentum of a physical system described by an algebraic special spacetime.

Example 11.1 In this example we shall explicitly construct a geodesic, shear-free, null congruence in flat spacetime called a **Kerr congruence**.

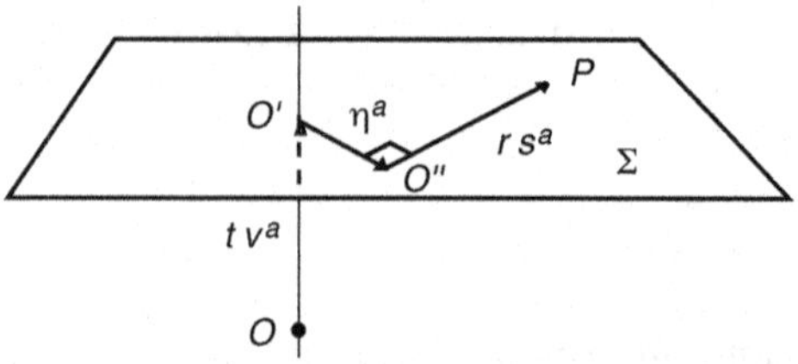

Figure 11.2. Point P is reached from O by traveling along the vectors tv^a, η^a, and rs^a.

Let v^a be a constant four-velocity vector, a^a a constant spacelike vector orthogonal to v^a, and O some point. The idea is to use the structure provided by these objects to construct a geodesic, shear-free, null congruence, or, in other words, a null vector field l^a such that $l^b\nabla_b l^a = 0$ and $\sigma = 0$.

Consider some three-plane Σ orthogonal to v^a. The world line through O in direction v^a will intersect Σ in a natural origin point O'. Given a unit spacelike vector s^a, orthogonal to v^a (i.e., $s^a s_a = -1$, $s^a v_a = 0$), the vector $\eta_c = \varepsilon_{abcd} v^a a^b s^d$ will be orthogonal to v^c, a^c, and s^c. In particular, being orthogonal to v^a, it will connect O' to some other point O'' in Σ. The set of all points O'' obtained in this way generates a toruslike, closed two-surface, which we call T. Given a point $P \in \Sigma$ but lying outside T, it is always possible to find a corresponding unique vector $s^a(P)$ and a corresponding positive number $r(P)$ such that $\eta^a(P) + r(P)s^a(P)$ connects O' to P (see Fig. 11.2). In three-vector notation, this can be written as

$$O'P = \boldsymbol{a} \times \mathbf{s}(P) + r(P)\mathbf{s}(P).$$

Any point outside T can thus be assigned a unique unit spacelike vector, $s^a(P)$, orthogonal to v^a, and a unique positive number $r(P)$ by means of this procedure. This gives a vector field and a scalar field, which we denote by s^a and r. The set of points on which s^a is constant forms a line tangent to s^a with parameter r. Thus $s^b\nabla_b s^a = 0$ and $s^b\nabla_b r = 1$. We also note that $s^a s_a = -1$ implies $s^a\nabla_b s_a = 0$.

The required null vector field is $l^a = v^a + s^a$. Since $s^b\nabla_b s^a = 0$ and $s^b\nabla_b r = 1$, we have $l^b\nabla_b l^a = 0$ and $l^b\nabla_b r = 0$, and thus l^a is tangent to a geodesic null congruence with parameter r.

In order to calculate the shear of this congruence we need the m^a-leg of a null tetrad. By completing v^a and s^a to form a right-handed ON-basis (v^a, s^a, e_2^a, e_3^a), we may write $m^a = (e_2^a + ie_3^a)/\sqrt{2}$. Note that this gives the required orthogonality relations, $l^a m_a = 0$, $m^a m_a = 0$, and $m^a\bar{m}_a = -1$. Furthermore, the right-handedness condition $\varepsilon_{abcd} v^a s^b e_2^c e_3^d = 1$ can easily be seen to imply

$$\varepsilon_{abcd} v^a s^b m^c \bar{m}^d = -i. \qquad (11.14)$$

From the definition of ρ and σ, we have $\rho = \bar{m}^b m^a \nabla_b l_a = \bar{m}^b m^a \nabla_b s_a$ and $\sigma = m^b m^a \nabla_b l_a = m^b m^a \nabla_b s_a$. These two relations, together with

$s^b\nabla_b s^a = 0$ and $s^a\nabla_b s_a = 0$, imply

$$\nabla_b s_a = \sigma\, \bar{m}_a \bar{m}_b + \bar{\sigma}\, m_a m_b + \bar{\rho} m_a \bar{m}_b + \rho \bar{m}_a\, m_b. \tag{11.15}$$

Furthermore, by expanding a^a in terms of the basis $(s^a, m^a, \bar{m}^a)$, we get

$$a^a = A s^a + \bar{B} m^a + B\bar{m}^a, \tag{11.16}$$

where

$$a\cos\theta = \boldsymbol{s}\cdot\boldsymbol{a} = -a^a s_a = A. \tag{11.17}$$

If x^a is the position vector field with respect to O, then

$$x_c = tv_c + \eta_c + rs_c = tv_c + \varepsilon_{abcd} v^a a^b s^d + rs_c.$$

Since $g_{ec} = \nabla_e x_c$, this gives

$$g_{ec} = \nabla_e t v_c + \varepsilon_{abcd} v^a a^b \nabla_e s^d + r\nabla_e s_c + \nabla_e r s_c.$$

Contracting this equation with $m^e m^c$ and using equations (11.14), (11.15), and (11.16), we get $0 = (Ai + r)\sigma = 0$, which implies $\sigma = 0$. The congruence is thus shear-free. Similarly, by contracting with $\bar{m}^e m^c$, we get $-1 = (Ai + r)\rho$ and hence

$$\rho = -\frac{1}{r + ia\cos\theta}, \tag{11.18}$$

where in the last step we used equation (11.17).

EXERCISES

11.1 Using the results of Exercise 9.5, prove

$$\hat{\rho} = \rho - \frac{D\Omega}{\Omega^3}.$$

11.2 Consider a null congruence formed by null cones emanating from the center-of-mass world line of a spherically symmetric mass distribution. Show that $\sigma = 0$ and that, in the vacuum region outside the distribution, there exists an affine parameter r such that $4\pi r^2 = A$, where A is the area of the $r = $ const cross sections of the null cones.

11.3 For a Kerr congruence, show that

$$\nabla_a r \nabla^a r = -\frac{r^2 + a^2}{r^2 + a^2\cos^2\theta}.$$

11.4 Show that the equations

$$D\rho = \rho^2 + \sigma\bar{\sigma} - \tfrac{1}{2} R_{ab}\, l^a\, l^b,$$
$$D\sigma = (\rho + \bar{\rho})\sigma + C_{abcd} l^a m^b l^c m^d$$

are conformally invariant.

11.5 For a surface-forming congruence in flat spacetime we have

$$D\rho = \rho^2 + \sigma\bar{\sigma},$$
$$D\sigma = 2\rho\sigma,$$
$$DA = -2\rho A.$$

Find the general solution of this system of equations.

11.6 Let σ_N be the shear of a null cone N in an algebraically special spacetime. Show that there exists a generator of N on which $\sigma_N = 0$.

12

Asymptotic Flatness and Symmetries

As we move away from an isolated body, its gravitational field decreases and tends to zero as we approach infinity. In a spacetime picture, where gravity is described by curvature, it would therefore seem entirely reasonable to model an isolated body on a spacetime that is, in some sense, asymptotically flat. If the body possesses some sort of symmetry – it might, for example, be an axisymmetric or spherically symmetric star – then it would also seem reasonable to model it on spacetime with the same type of symmetry. But what exactly do we mean by asymptotic flatness, and what do we mean by a spacetime symmetry? In this chapter we shall attempt to answer these two questions.

12.1 Asymptotically Flat Spacetimes

In order to use general relativity to study the gravitational field of an isolated body, such as a star, it is necessary to have some well-defined notion of asymptotic flatness. An asymptotically flat spacetime represents the idealized situation of a gravitating body that, to all intents and purposes, is totally isolated from the rest of the universe by virtue of its great distance from all other bodies. As we move away from such an object, its gravitational field should decrease, and thus we expect that spacetime should become flat at asymptotic distances. Of course, no physical system truly can be isolated from the rest of the universe, but for a system such as a star the gravitational influence of all other matter is so slight that it is entirely reasonable to consider it as being totally isolated, essentially a single system in an otherwise empty universe.

As a preliminary definition, we say that a spacetime is asymptotically flat if there exists a coordinate system (x^0, x^1, x^2, x^3) such that the corresponding components of the metric satisfy $g_{\alpha\beta} = \eta_{\alpha\beta} + O(1/r)$ as $r \to \infty$ along any spacelike or null direction, where $r = \sqrt{(x^1)^2 + (x^2)^2 + (x^3)^2}$. We also demand that $R_{ab} = 0$ for sufficiently large r, since T_{ab} and hence (by Einstein's equation) R_{ab} must vanish outside the world tube of the source.

Although this definition is adequate for most of our purposes and has the virtue of being simple to understand, it is defective in a number of respects. First, it is nongeometric in that it is coordinate-dependent. Apart from purely aesthetic considerations, this means that it is difficult to work with, since the coordinate invariance of all statements must be carefully checked, and this involves many troublesome issues concerning the interchange of limits and derivatives. Second, it demands the existence of a global coordinate system, which imposes a severe restriction on the spacetime and one that has nothing to do with the notion of asymptotic flatness. This difficulty can easily be overcome by restricting the domain of the coordinate functions, but this has the drawback of leading to a complicated and unwieldy definition. We shall shortly replace it with a more geometric definition, but for now we shall use it to obtain a result that we shall find useful later on.

The existence of a radial coordinate function r satisfying the above properties is important, as it tells us how to approach infinity and distinguishes curves that escape to infinity from these that don't: a spacelike or null curve, $\lambda \to p(\lambda)$, escapes to infinity if $r(p(\lambda)) \to \infty$ as $\lambda \to \infty$. If, on the other hand, $r(p(\lambda))$ remains bounded, then the curve does not escape to infinity but always remains at a finite distance from the source. We shall encounter such situations when we come to consider black holes in a later chapter.

The above definition does not, of course, determine a unique radial function, but all radial functions satisfying the definition have the following asymptotic property:

$$\nabla_a r \nabla^a r \to -1 \tag{12.1}$$

as infinity is approached in any direction.

Let us now proceed toward a more geometric definition of asymptotic flatness. This is by no means an easy task and constitutes a whole branch of relativity in itself, with several outstanding problems still unsolved. The first main difficulty is to obtain a definition that fully captures the intuitive notion of an isolated system within a spacetime context, and yet allows sufficiently many physically interesting realizations. We do not want, for example, a definition that does not admit spacetimes having an asymptotic flux of gravitational radiation. This would certainly be the case for too strong a definition. A second difficulty is that infinity can be approached in several physically distinct directions: in a timelike direction, either toward the past or toward the future; in a null direction, again toward either the past or future; and in a spacelike direction. There are thus several possible types of infinity for a spacetime: future and past timelike infinity, future and past null infinity, and spacelike infinity. Here we consider just one of these, namely, future null infinity.

The basic idea behind the definition [originally due to Penrose (1968)] is to add extra points, representing points at infinity, to the spacetime manifold M – a familiar idea in projective geometry, where parallel lines intersect at a point at infinity. This results in a larger manifold $\hat{M} = M \cup \mathcal{I}^+$

with boundary $\mathcal{I}^+$ representing points at future null infinity. The points of $\mathcal{I}^+$ are then brought in to a finite "distance" with respect to a new, nonphysical, unit of "length" based on a conformally related metric, $\hat{g} = \Omega^2 g$, where the conformal factor Ω vanishes on $\mathcal{I}^+$ (i.e. infinite stretching at infinity). If such a procedure can be carried out and results in a (nonphysical) spacetime $(\hat{M}, \hat{g})$, where $\hat{g}$ is well defined on $\mathcal{I}^+$, the physical spacetime (M, g) is said to be asymptotically flat at future null infinity.

Before writing down the definition, let us first illustrate the basic ideas by considering a flat spacetime M with the metric

$$g = dt^2 - dr^2 - r^2(d\theta^2 + \sin^2\theta\, d\varphi^2)$$

expressed in spherical coordinates. In terms of a retarded spherical coordinate system (u, r, θ, φ), where $u = t - r$, the metric takes the form

$$g = du^2 + 2\, du\, dr - r^2(d\theta^2 + \sin^2\theta\, d\varphi^2).$$

As can easily be seen from the form of the metric, the level surfaces of u are future null cones, and r serves as an affine parameter along their null rays, that is,

$$l^a l_a = 0 \quad \text{and} \quad l^a \nabla_a r = 1, \tag{12.2}$$

where $l_a = \nabla_a u$ is future-pointing. For future reference we also note that

$$\nabla_a r \nabla^a r = -1, \tag{12.3}$$

which we have already noted is a characteristic property of a radial function. Future null infinity can thus be approached along a null ray by keeping u, θ, and φ fixed and allowing r to tend to infinity.

In order to give points at future null infinity a finite coordinate value, we introduce a new coordinate system $(u, \Omega, \theta, \varphi)$, where $\Omega = r^{-1}$, in terms of which the metric becomes

$$g = \Omega^{-2}(\Omega^2 du^2 - 2\, du\, d\Omega - d\theta^2 - \sin^2\theta\, d\varphi^2).$$

By means of the function Ω we are able to extend the spacetime manifold M, which we identify with the set of all coordinate points $(u, \Omega, \theta, \varphi)$ where $\Omega > 0$, to a larger manifold $\hat{M}$, which we identify with the set of all coordinate points $(u, \Omega, \theta, \varphi)$ where $\Omega \geq 0$. The extended space, $\hat{M} = M \cup \mathcal{I}^+$, has a boundary $\mathcal{I}^+$ with topology $\mathbb{R} \times S^2$ on which $\Omega = 0$. We refer to $\mathcal{I}^+$ as **future null infinity**.

Even though the coordinate system $(u, \Omega, \theta, \varphi)$ is well defined on $\mathcal{I}^+$, the form of the physical metric g in terms of these coordinates blows up as we approach $\mathcal{I}^+$, due to the factor Ω^{-2}. This is, of course, what one might expect, since an infinite amount of coordinate stretching is needed to give points of $\mathcal{I}^+$ a finite coordinate value. However, the conformally rescaled metric

$$\hat{g} = \Omega^2 g = \Omega^2\, du^2 - 2\, du\, d\Omega - d\theta^2 - \sin^2\theta\, d\varphi^2 \tag{12.4}$$

admits a smooth extension to $\mathcal{I}^+$ giving $(\hat{M}, \hat{g})$ the structure of a (nonphysical) spacetime with a boundary $\mathcal{I}^+$ on which $\Omega = 0$ and $\nabla_a\Omega \neq 0$ [this is because the coordinate system $(u, \Omega, \theta, \varphi)$ is well defined on $\mathcal{I}^+$].

Since we now have two metrics, g_{ab} and $\hat{g}_{ab} = \Omega^2 g_{ab}$ (note that $\hat{g}^{ab} = \Omega^{-2} g^{ab}$), we also have two corresponding connections, ∇_a and $\hat{\nabla}_a$. These will differ when acting on tensors but coincide when acting on scalars (we recall that the gradient operator is metric-independent); for example, $\nabla_a\Omega = \hat{\nabla}_a\Omega$. In order to cope with this duplication of objects, we employ the convention of using g_{ab} and g^{ab} to raise and lower indices of unhatted quantities, and $\hat{g}_{ab}$ and $\hat{g}^{ab}$ for hatted quantities. Thus, for example, $\hat{\nabla}^a\Omega = \hat{g}^{ab}\hat{\nabla}_b$ and $\nabla^a\Omega = g^{ab}\nabla_b\Omega$.

From the form of the metric $\hat{g}$, we see that

$$\hat{\nabla}_a\Omega\hat{\nabla}^a\Omega = \Omega^2 \quad \text{and} \quad \hat{\nabla}^a\Omega\hat{\nabla}_a u = -1.$$

The first equation implies that $\hat{\nabla}_a\Omega\hat{\nabla}^a\Omega = 0$ on $\mathcal{I}^+$ and hence $\mathcal{I}^+$ is a null boundary with tangent vector $\hat{n}^a = -\hat{\nabla}^a\Omega$. Taking a derivative of the first equation, we also have

$$\hat{\nabla}_b(\hat{\nabla}_a\Omega\hat{\nabla}^a\Omega) = 2\hat{\nabla}^a\Omega\hat{\nabla}_b\hat{\nabla}_a\Omega = 2\hat{\nabla}^a\Omega\hat{\nabla}_a\hat{\nabla}_b\Omega = 2\Omega\hat{\nabla}_b\Omega,$$

and hence $\hat{n}^a\hat{\nabla}_a\hat{n}_b = 0$ on $\mathcal{I}^+$. In other words, $\hat{n}^a$ is an *affine* tangent vector to the null generators of $\mathcal{I}^+$. The second equation implies that $\hat{n}^a\hat{\nabla}_a u = 1$ on $\mathcal{I}^+$, and hence u is an *affine* parameter along the generators. Since $l_a = \nabla_a u = \hat{\nabla}_a u$ is future-pointing, $\hat{n}^a$ must also be future-pointing on $\mathcal{I}^+$.

The key idea behind all this is that flat spacetime allows a conformal completion by virtue of its asymptotic properties. This suggests that we define a curved spacetime to be asymptotically flat at future null infinity if it allows a similar completion to be performed. By sifting out the geometric, coordinate-independent content of the completion of flat spacetime, we arrive at the following definition of an asymptotically flat curved spacetime.

Definition 12.1 A spacetime (M, g_{ab}) is asymptotically flat at future null infinity if M can be extended to a larger manifold $\hat{M} = M \cup \mathcal{I}^+$ with boundary $\mathcal{I}^+$ of topology $\mathbb{R} \times S^2$ such that:

(i) There exists a function Ω on $\hat{M}$ such that $\Omega > 0$ on M and that $\Omega = 0$ and $\hat{\nabla}_a\Omega \neq 0$ on $\mathcal{I}^+$. This simply states that $\mathcal{I}^+$ is a proper boundary of $\hat{M}$.

(ii) There exists a metric, $\hat{g}_{ab}$, on $\hat{M}$ such that $\hat{g}_{ab} = \Omega^2 g_{ab}$ on M.

(iii) On $\mathcal{I}^+$, $\hat{n}_a = -\hat{\nabla}_a\Omega$ is null (i.e., $\mathcal{I}^+$ is a null boundary) and future-pointing. That $\hat{n}_a$ is future-pointing means that we are dealing with future rather than past null infinity.

(iv) There exists a neighborhood of $\mathcal{I}^+$ on which $R_{ab} = 0$. This ensures that the physical Ricci curvature R_{ab} (and hence the energy momentum tensor) vanishes in the asymptotic region far away from the source of the gravitational field.

It should be emphasized that all statements involving derivatives on $\hat{M}$ refer to the smooth manifold structure of $\hat{M}$.

Allowing for a few minor topological difficulties, the above definition can be shown to be equivalent to the previous, preliminary definition (at least where future null directions are concerned), but has the distinct advantage of being expressed in a purely geometric form. One of its most important features is that it relies on a conformal transformation of the metric, and many spacetime notions are invariant under such a transformation. For example, as we have already seen in Section 10.1, the notion of a null geodesic is invariant under a conformal transformation, and the same can also be shown to be true for massless fields (e.g. a Maxwell field). A discussion of massless particles or fields can thus take place just as well in $\hat{M}$ as in the actual physical spacetime M. Needless to say, this leads to many conceptual and mathematical advantages.

Perhaps the greatest virtue of the conformal definition of $\mathcal{I}^+$ is that it provides a basis for the rigorous discussion of gravitational radiation, without recourse to dubious approximation schemes. Indeed, without $\mathcal{I}^+$ it would be difficult to assign a precise meaning to the very notion of gravitational radiation. This is because some sort of rigid background is required to distinguish "radiative" curvature (gravity) from the ordinary "Coulomb" variety. Future null infinity, $\mathcal{I}^+$, provides precisely the sort of background required, and this has led to a much deeper theoretical understanding of the nature and properties of radiation within the context of general relativity.

A conformal factor Ω satisfying the conditions of the definition is not determined uniquely, but only up to $\Omega \to \omega\Omega$, where ω is nonzero at all points of $\mathcal{I}^+$. Since the physical metric g_{ab} is fixed and $\hat{g}_{ab} = \Omega^2 g_{ab}$, we have $\hat{g}_{ab} \to \omega^2 \hat{g}_{ab}$ and $\hat{g}^{ab} \to \omega^{-2}\hat{g}^{ab}$. Let us now use this conformal freedom to obtain a more rigid structure on $\mathcal{I}^+$, similar to that which we obtained in the case of flat spacetime. Since we shall have recourse to evaluate many quantities on $\mathcal{I}^+$ we define $x \mathrel{\hat{=}} y$ to mean $x = y$ when evaluated on $\mathcal{I}^+$. Thus, for example, $\Omega \mathrel{\hat{=}} 0$ and $\hat{\nabla}_a\Omega\hat{\nabla}^a\Omega \mathrel{\hat{=}} 0$.

Since $\hat{\nabla}_a\Omega\hat{\nabla}^a\Omega \mathrel{\hat{=}} 0$, smoothness implies that $\hat{\nabla}_a\Omega\hat{\nabla}^a\Omega/\Omega \mathrel{\hat{=}} f$ is a well-defined function on $\mathcal{I}^+$. A simple calculation shows that, under our conformal transformation, f transforms like

$$f \to -\omega^{-2}\hat{n}^a\hat{\nabla}_a\omega + \omega^{-1} f.$$

We may thus choose ω on $\mathcal{I}^+$ so that $f = 0$. This restricts our conformal freedom up to an ω satisfying $\hat{n}^a \hat{\nabla}_a \omega := 0$, that is, ω is constant along the generators of $\mathcal{I}^+$.

To proceed further we now need to use the fact that $R_{ab} = 0$ near $\mathcal{I}^+$. It can be shown that R_{ab} is related to $\hat{R}_{ab}$ (the Ricci curvature of $\hat{g}_{ab}$) by the equation

$$R_{ab} = \hat{R}_{ab} - 2\Omega^{-1}\hat{\nabla}_a\hat{\nabla}_b\Omega - \hat{g}_{ab}(\Omega^{-1}\hat{\nabla}_c\hat{\nabla}^c\Omega - 3\Omega^{-2}\hat{\nabla}_c\Omega\hat{\nabla}^c\Omega). \qquad (12.5)$$

Since $R_{ab} := 0$, $\Omega^{-1}\hat{\nabla}_a\Omega\hat{\nabla}^a\Omega := 0$ and $\hat{R}_{ab}$ is well defined on $\mathcal{I}^+$, this gives

$$2\hat{\nabla}_a\hat{\nabla}_b\Omega + \hat{g}_{ab}\hat{\nabla}_c\hat{\nabla}^c\Omega := 0.$$

Contracting on a and b gives $\hat{\nabla}_c\hat{\nabla}^c\Omega := 0$ and hence $\hat{\nabla}_a\hat{\nabla}_b\Omega := 0$. This implies, trivially, that $\hat{n}^a$ is an *affine* tangent vector to the generators of $\mathcal{I}^+$ and that $\mathcal{I}^+$ is divergence- and shear-free.

Let us now introduce coordinate functions, θ and φ, on $\mathcal{I}^+$ that are constant along its generators. If Σ is some spacelike S^2, cross section of $\mathcal{I}^+$ then (θ, φ) serves as a coordinate system on Σ. Since Σ has topology S^2, a result from differential geometry shows that (θ, φ) may be chosen such that the induced metric on Σ has the form $V^2(d\theta^2 + \sin^2\theta\, d\varphi^2)$; that is, it is conformal to a metric (round) two-sphere. By making use of the remaining conformal freedom, we now set $V = 1$, thus making Σ into a metric two-sphere with intrinsic metric $d\theta^2 + \sin^2\theta\, d\varphi^2$ and area element $dS = \sin\theta\, d\theta\, d\varphi$. Note that this implies a total area equal to 4π. Since a metric two-sphere with area 4π would have unit radius when embedded in Euclidean three-space, we shall call such an object a **unit sphere**.

Let Σ' be some other spacelike, S^2 cross section of $\mathcal{I}^+$. Since (θ, φ) also serves as a coordinate system on Σ', we have a natural correspondence between points of Σ and points of Σ', where points are in correspondence if they have the same coordinate values, or, equivalently, if they are connected by a generator. Since $\mathcal{I}^+$ is divergence-free, our knowledge of null congruences tells us that an area element does not change under this mapping, and hence the area element of Σ' is also given by $dS = \sin\theta\, d\theta\, d\varphi$. This implies that the intrinsic metric of Σ' is also given by $d\theta^2 + \sin^2\theta\; d\varphi^2$ and hence Σ' is also a unit sphere.

To sum up, we have shown that there exists a conformal scaling of $\mathcal{I}^+$ such that:

(i) $\hat{n}^a = -\hat{\nabla}^a\Omega$ is an affine tangent vector to the null generators of $\mathcal{I}^+$, that is, $\hat{n}^a\hat{\nabla}_a\hat{n}^b := 0$.
(ii) $\mathcal{I}^+$ is shear- and divergence-free.
(iii) All spacelike S^2 cross sections of $\mathcal{I}^+$ are unit two-spheres.

Such a scaling on $\mathcal{I}^+$ is sometimes called a **Bondi scaling**.

By introducing a function u that serves as an affine parameter along the generators of $\mathcal{I}^+$ in that $\hat{n}^a\hat{\nabla}_a u := 1$, we obtain a coordinate system

$(u, \Omega, \theta, \varphi)$ that is well defined in a neighborhood of $\mathcal{I}^+$. Furthermore, if $\Omega = r^{-1}$, then (u, r, θ, φ) serves as an asymptotic coordinate system on M. This is sometimes called a **Bondi coordinate system**.

There are several ways of extending the coordinate functions of a Bondi system into the interior of M. The most common is as follows. First, we demand that the level surfaces of u be null, that is, we choose u such that $l_a = \nabla_a u$ is a null vector. Second, we choose r such that $l^a \nabla_a r = 1$, that is, we choose r such that it is an affine parameter along the null rays of l^a. Third, we choose the angular coordinates such that they are constant along the null rays of l^a, that is, $l^a \nabla_a \theta = l^a \nabla_a \varphi = 0$. The resulting coordinate system can be shown to be a polar version of a Cartesian system (x^0, x^1, x^2, x^3) satisfying the conditions our preliminary definition of asymptotic flatness – at least for future null directions.

Since a null geodesic is a conformally invariant notion, the conformal definition of asymptotic flatness allows a particularly natural setting for discussing the asymptotic behavior of null geodesic congruences. We say that a null geodesic congruence is asymptotically well behaved if, in the $\hat{M}$ picture, each curve intersects $\mathcal{I}^+$ in a point and no two distinct curves intersect $\mathcal{I}^+$ in a single point. A null cone congruence in flat spacetime is, for example, asymptotically well behaved. We now show that an asymptotically well behaved congruence, together with a particular choice of Bondi scaling on $\mathcal{I}^+$, leads to a unique conformal factor, Ω, such that $r = 1/\Omega$ is an affine parameter in the M picture.

Let $\hat{l}^a$ be an affine tangent vector to the congruence. Given a Bondi scaling, the remaining conformal freedom is given by $\hat{g}_{ab} \to \omega^2 \hat{g}_{ab}$ and $\Omega \to \omega\Omega$ where ω=1 on $\mathcal{I}^+$. In particular $\hat{n}_a := -\hat{\nabla}_a \Omega$ is unique on $\mathcal{I}^+$. We can thus fix $\hat{l}^a$ on $\mathcal{I}^+$ by demanding $\hat{l}^a \hat{n}_a = 1$, in other words, $\hat{D}\Omega := -1$, where $\hat{D} = \hat{l}^a \hat{\nabla}_a$. This fixes $\hat{l}^a$ up to $\hat{l}^a \to \omega^{-2} \hat{l}^a$. Let us now use up most of this freedom by demanding $\hat{D}\Omega := -1$ at *all* points. This determines ω up to a solution of

$$\Omega \hat{D}\omega + \omega = \omega^2$$

where $\omega = 1$ on $\mathcal{I}^+$. Since $\Omega = 0$ on $\mathcal{I}^+$, this allows $\hat{D}\omega$ to be chosen freely on $\mathcal{I}^+$. However, once this choice has been made, ω is unique and we have a unique conformal factor Ω.

We choose $\hat{D}\omega$ on $\mathcal{I}^+$ by making use of the divergence function $\hat{\rho}$. From equation (11.12) we see that $\hat{\rho} \to \hat{\rho} + \hat{D}\omega$ on $\mathcal{I}^+$. We may thus fix $\hat{D}\omega$ by demanding that the real part of $\hat{\rho}$ vanish on $\mathcal{I}^+$. We now have a unique conformal factor and a unique affine tangent vector such that $\hat{D}\Omega = -1$.

Let us now transfer to the physical spacetime M and define a radial function r by $\Omega = 1/r$. By Theorem 11.2 we see that $l^a = \Omega^2 \hat{l}^a$ is an affine tangent vector with respect to the physical metric. Furthermore, $\hat{D}\Omega = \Omega^{-2} D\Omega = -1$ implies $Dr = 1$. In other words, r is an affine parameter.

Let us now investigate the asymptotic behavior of ρ and σ. Using Theorem 11.2 and equation (11.12), we have

$$\rho = -r^{-1} + \hat{\rho} r^{-2} \quad \text{and} \quad \sigma = \hat{\sigma} r^{-2}. \tag{12.6}$$

These two equations can be expressed in a slightly different form by introducing functions σ^0 and ρ^0 defined uniquely by $D\sigma^0 = D\rho^0 = 0$ and $\sigma^0 = \hat{\sigma}$ and $\rho^0 = \hat{\rho}$ on $\mathcal{I}^+$. Note that, by construction, the real part of $\hat{\rho}$ is zero on $\mathcal{I}^+$ and hence ρ^0 is purely imaginary. Using these function we obtain the important asymptotic equations

$$\rho = -r^{-1} + \rho^0 r^{-2} + O(r^{-3}) \quad \text{and} \quad \sigma = \sigma^0 r^{-2} + O(r^{-3}). \tag{12.7}$$

If the congruence is shear-free ($\hat{\sigma} = \sigma = 0$), then equation (11.13) gives $\rho = -1/(r + i\Sigma)$ and, on comparing this with equation (12.7), we see that $\rho^0 = i\Sigma$ or, equivalently, $\hat{\rho} = i\Sigma$ on $\mathcal{I}^+$.

12.2 Killing Fields and Stationary Spacetimes

A spacetime may be said to possess a symmetry if there exists a coordinate system, (u, r, θ, φ) say, such that the components $g_{\alpha\beta}$ of the metric are independent of one or more of the coordinate functions. For example, if the spacetime corresponds to a stationary star, then we expect there to exist, at least locally, a coordinate system (u, r, θ, φ) such that $g_{\alpha\beta}$ is independent of the "time" coordinate u. In this case we say that spacetime possesses a timelike symmetry and is stationary. If, in addition, the star is axisymmetric, then we expect there to exist a coordinate system (u, r, θ, φ) such that $g_{\alpha\beta}$ is independent of *both* u and φ. In this case we say the spacetime possesses two compatible symmetries and is stationary and axisymmetric.

The great drawback of this definition of a spacetime symmetry is that it is coordinate-dependent and nongeometric. Fortunately, it can be expressed in a geometric fashion by the aid of our old friend, the Lie derivative. Suppose that $g_{\alpha\beta}$ are independent of u, that is, $\partial_u g_{\alpha\beta} = 0$. Then, as we saw in Section 7.2, this implies $\mathcal{L}_\xi g_{ab} = 0$, where $\xi^a = \partial_u^a$. Conversely, if there exists a vector field ξ^a such that $\mathcal{L}_\xi g_{ab} = 0$, then there exists, at least locally, a coordinate system, (u, r, θ, φ) say, such that $\xi^a = \partial_u^a$ and where $\partial_u g_{\alpha\beta} = 0$. This leads to a more geometric definition of a spacetime symmetry: a spacetime possesses a symmetry if it admits a vector field ξ^a, called a **Killing vector**, such that $\mathcal{L}_\xi g_{ab} = 0$. By equation (7.13) we see that this is equivalent to the **Killing equation**,

$$\nabla_{(a}\xi_{b)} = 0. \tag{12.8}$$

The streamlines of a Killing vector are called its **orbits** and give curves along which the spacetime environment remains constant.

Reverting to our coordinate definition, if we have two compatible symmetries in that the components $g_{\alpha\beta}$ are independent of u and φ, then we have two Killing vectors, $\xi^a = \partial_u^a$ and $\chi^a = \partial_\varphi^a$. Furthermore, the components of ∂_φ^a [namely, (0, 0, 01)] are independent of u, and hence $\mathcal{L}_\xi \chi^a = [\xi, \chi]^a = 0$. Conversely, if there exist two Killing vectors ξ^a and χ^a such that $[\xi, \chi]^a = 0$, then there exists, at least locally, a coordinate system, (u, r, θ, φ) say, such that $\xi^a = \partial_u^a$ and $\chi^a = \partial_\varphi^a$ and that $\partial_u g_{\alpha\beta} = \partial_\varphi g_{\alpha\beta} = 0$.

This all leads to the following definition:

Definition 12.2

(i) A spacetime is **stationary** if it admits a Killing vector ξ^a that is asymptotically timelike.

(ii) A spacetime is **axisymmetric** if it admits a Killing vector χ^a with spacelike, closed orbits.

(iii) A spacetime is axisymmetric and stationary if it satisfies the previous two conditions together with the compatibility condition, $[\xi, \chi] = 0$.

Physically, a stationary star can either be rotating or nonrotating. A nonrotating star corresponds to a special type of stationarity where there are no preferred lateral directions (if the earth were nonrotating, there would be little distinction between east and west). This will be the case if the Killing field ξ^a is orthogonal to some spacelike three-surface. A stationary spacetime whose Killing vector satisfies this additional condition is said to be **static**.

If it is to describe an object such as a rotating star, a stationary, axisymmetric spacetime should also satisfy certain discrete symmetries. If, for example, we take a motion picture of a rotating star and run the film backwards, what do we see? We see exactly the same physical situation except that the star appears to be rotating in the opposite direction. If, however, we turn the film over, so right becomes left and left becomes right, and again run the film backwards, we see exactly the same physical situation with the star rotating in the correct direction. In other words, a physical system such as a rotating star is *TP* (time–parity) symmetric. In terms of a spacetime description, *TP* symmetry can be defined as follows. Let S represent the sum total of all geometric structure that arises from the metric without reference to any arbitrary choice of timelike or spacelike orientation. Then our spacetime is *TP*-symmetric if (ξ^a, χ^a) and $(-\xi^a, -\chi^a)$ are both related to S in exactly the same way. For example, if our spacetime is *TP*-symmetric and S contains a vector l_a such that $l_a\xi^a = X$ and $l_a\chi^a = Y$, then S must also contain a corresponding vector, l'_a, such that $l'_a\xi^a = -X$ and $l'_a\chi^a = -Y$.

Properties of Killing Fields

Throughout the rest of this book we shall have frequent occasion to use Killing vector fields. We therefore devote this section to investigating and proving some of their most important properties.

Perhaps the most frequently used property of a Killing field is this:

Theorem 12.1 *Let ξ^a be a Killing field and γ a geodesic with (affine) tangent vector u^a, i.e. $u^b\nabla_b u^a = 0$. Then $\xi_a u^a$ is constant along γ.*

Proof We have

$$u^b\nabla_b(\xi_a u^a) = u^b u^a \nabla_b \xi_a + \xi_a u^b \nabla_b u^a = 0,$$

where the first term on the right-hand side is zero by the Killing equation, and the second term is zero by the geodesic condition. □

We shall use this theorem shortly to investigate red shifts in the presence of a gravitational field.

Since $\nabla_{(a}\xi_{b)} = 0$,

$$f_{ab} = \nabla_a \xi_b \tag{12.9}$$

is antisymmetric and therefore a 2-form. This 2-form in turn defines a conserved current j_a by

$$j_b = \nabla^a f_{ab}. \tag{12.10}$$

Our next theorem relates this current to the Ricci curvature and hence, via the Einstein equation, to the matter content of spacetime.

Theorem 12.2

$$j_a = R_{ab}\xi^b. \tag{12.11}$$

Proof By the definition of the curvature tensor, we have

$$\nabla_a\nabla_b\xi_c - \nabla_b\nabla_a\xi_c = -R_{abc}{}^d\xi_d,$$

which, on using Killing's equation, gives

$$\nabla_a\nabla_b\xi_c + \nabla_b\nabla_c\xi_a = -R_{abc}{}^d\xi_d.$$

Furthermore, by permuting the indices and using equation (9.27), namely

$$R_{abcd} + R_{adbc} + R_{acdb} = 0,$$

we find that

$$\begin{aligned} 2\nabla_b\nabla_c\xi_a &= -(R_{abc}{}^d + R_{bca}{}^d - R_{cab}{}^d)\xi_d \\ &= 2R_{cab}{}^d\xi_d. \end{aligned}$$

Finally, on contracting this equation, we get the required result,

$$\nabla^c\nabla_c\xi_a = R_{ca}{}^{cd}\xi_d = R_a{}^d\xi_d. \qquad \square \tag{12.12}$$

As we shall see shortly, this theorem allows mass and angular momentum to be well defined for a stationary, axisymmetric spacetime.

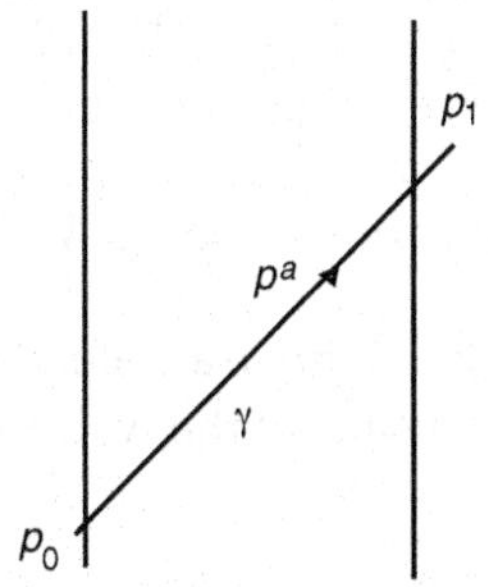

Figure 12.1. A photon γ of four-momentum p^a is transmitted from p_0 and received at p_1.

The Red-Shift Factor

Given an timelike Killing vector ξ^a, we define the **red-shift factor** as the positive function V given by

$$V^2 = \xi_a \xi^a, \tag{12.13}$$

where, for convenience, we choose ξ^a such that $V \to 1$ as we approach infinity. When evaluated at some point p, $v^a = V^{-1}\xi^a$ gives the four-velocity of a stationary observer occupying p.

Consider now two points p_0 and p_1 connected by a null geodesic γ with p_1 lying to the future of p_0 (Fig. 12.1). If γ is the null ray of a photon with four-momentum p_a, then its energy–frequency with respect to the observer at p_0 is

$$\omega_0 = p_a v^a(p_0) = V^{-1}(p_0) p_a \xi^a(p_0),$$

and, similarly, its energy–frequency with respect to the observer at p_1 is

$$\omega_1 = p_a v^a(p_1) = V^{-1}(p_1) p_a \xi^a(p_1).$$

But, by Theorem 12.1, we have $p_a \xi^a(p_0) = p_a \xi^a(p_1)$ and hence

$$\frac{\omega_1}{\omega_0} = \frac{V(p_0)}{V(p_1)}. \tag{12.14}$$

In particular, if p_1 is a point at infinity, then $V(p_1) = 1$ and we get

$$\omega_1 = \omega_0 V(p_0). \tag{12.15}$$

Thus, if $V \leq 1$, a photon will suffer a gravitational red shift on its journey from the interior out to infinity. We shall now show that $V \leq 1$ is indeed the case for a physically reasonable system such as a star.

By the Killing equation we have $\nabla_a \xi_b = -\nabla_b \xi_a$ and hence

$$2\xi^b \nabla_b \xi_a = -2\xi^b \nabla_a \xi_b = -\nabla_a V^2 = -2V \nabla_a V. \tag{12.16}$$

Using this equation together with $\xi^a \nabla_a V = 0$, we get

$$v^b \nabla_b v_a = -V^{-1} \nabla_a V, . \tag{12.17}$$

where $v^a = V^{-1}\xi^a$. The vector $v^b\nabla_b v_a$ is acceleration of a stationary observer (e.g., someone sitting under an apple tree), and thus

$$a_a = -v^b\nabla_b v_a = V^{-1}\nabla_a V$$

is the *relative* acceleration of a freely moving particle (e.g. a falling apple). By means of the well-verified observation that apples fall rather than rise, we see that

$$\boldsymbol{r}\cdot\boldsymbol{a} = -r^a a_a = -r^a V^{-1}\nabla_a V < 0$$

where r^a is an outward-pointing spacelike vector and $\boldsymbol{r}\cdot\boldsymbol{a}$ is the three-space (positive-definite) scalar product of a^a and r^a. The function V thus increases in an outward radial direction, and since $V = 1$ at infinity, $V < 1$ in the interior.

12.3 Kerr Spacetime

We now have at our disposal six natural, geometric conditions that can be imposed on a spacetime: asymptotic flatness, stationarity, axisymmetry, algebraic specialness, *TP* invariance, and the vacuum condition $R_{ab} = 0$. If all six conditions are satisfied, we get what is referred to as a **Kerr spacetime**. In this section we shall simply use these conditions to investigate certain important geometric properties of a Kerr spacetime and leave the full treatment till later.

Let us first bring the conformal definition of asymptotic flatness more firmly into the picture. As we have seen, an asymptotically flat spacetime admits a special class of coordinate systems, called Bondi systems. These are about the closest one can come to null polar coordinate systems in flat spacetime. For an asymptotically flat spacetime, which is also stationary and axisymmetric, there will exist a Bondi system (u, r, θ, φ) such that the components of the metric are independent of u and φ. There will, of course, exist other, less rigidly defined coordinate systems that exhibit the same symmetries, but the advantage of using a Bondi system in this context is that the corresponding Killing vectors, $\xi^a = \partial_u^a$ and $\chi^a = \partial_\varphi^a$, automatically have a natural asymptotic normalization in that $\xi_a\xi^a \to 1$, and $r^{-2}\chi_a\chi^a \to -\sin^2\theta$ near $\mathcal{I}^+$. Writing $\xi^a = \hat{\xi}^a$, $\chi^a = \hat{\chi}^a$, $\Omega = r^{-1}$, and $\hat{g}_{ab} = \Omega^2 g_{ab}$, we also note the following:

(i) $\hat{\xi}^a\hat{\nabla}_a u = \hat{\chi}^a\hat{\nabla}_a\varphi = 1$ and

$$\hat{\xi}^a\hat{\nabla}_a\Omega = \hat{\xi}^a\hat{\nabla}_a\theta = \hat{\xi}^a\hat{\nabla}_a\varphi = \hat{\chi}^a\hat{\nabla}_a u = \hat{\chi}^a\hat{\nabla}_a\Omega = \hat{\chi}^a\hat{\nabla}_a\theta = 0.$$

(ii) $\hat{\xi}_a\hat{\xi}^a = 0$ on $\mathcal{I}^+$, and hence, being null, $\hat{\xi}^a$ points along the generators of $\mathcal{I}^+$.

(iii) $\hat{\chi}_a\hat{\chi}^a = -\sin^2\theta$ on $\mathcal{I}^+$.

(iv) On $\mathcal{I}^+$, $\hat{\xi}^a\hat{\nabla}_a u = -\hat{\nabla}^a u\hat{\nabla}_a\Omega = 1$. Therefore, since $\hat{\nabla}_a\Omega$ and $\hat{\xi}_a$ both point along a generator, $\hat{\xi}_a = -\hat{\nabla}_a\Omega$.

This leads to the following theorem:

Theorem 12.3 *Let M be an asymptotically flat, stationary, axisymmetric spacetime with normalized Killing fields ξ^a and χ^a. Then there exists a conformal completion $\hat{M}$ of M, with a unique Bondi scaling on $\mathcal{I}^+$, and two Killing fields $\hat{\xi}^a$ and $\hat{\chi}^a$ such that:*

(i) $\hat{\xi}^a = \xi^a$ *and* $\hat{\chi}^a = \chi^a$ *on M.*
(ii) $\hat{\xi}^a\hat{\nabla}_a\Omega = \hat{\chi}^a\hat{\nabla}_a\Omega = 0$.
(iii) $\hat{\xi}_a = -\hat{\nabla}_a\Omega$ *and* $\Omega^{-2}\hat{\xi}_a\hat{\xi}^a = 1$ *on* $\mathcal{I}^+$.
(iv) *There exists a function θ, unique on $\mathcal{I}^+$, such that* $\hat{\chi}_a\hat{\chi}^a = -\sin^2\theta$ *on* $\mathcal{I}^+$.

An important feature of this theorem is that it highlights the fact that ξ^a fixes a *unique* Bondi scaling on $\mathcal{I}^+$, and that χ^a determines a *unique* **latitudinal angle** θ on $\mathcal{I}^+$. This means that any geometrically defined scalar on $\mathcal{I}^+$ can be expressed as some unique function of θ. In the interior, any geometrically defined scalar can be expressed as *some* function of θ and Ω (or r), but since we have considerable freedom in choosing θ and Ω in the interior, these functions will not be unique. This should be contrasted with the corresponding case in Newtonian theory, where the presence of an axisymmetric body allows a unique latitudinal angle $\theta(P)$ and a unique radial value $r(P)$ to be assigned to any point P (Fig. 12.2).

If we wish the same to be true in general relativity, then we must find some way of extending r and θ into the interior. This can be done for θ by introducing a null congruence (asymptotically well defined) and demanding that θ be constant along its curves, that is, $l^a\nabla_a\theta = 0$. Furthermore, as we saw in Section 11.1, our Bondi scaling fixes the affine tangent vector l^a and the affine parameter r. Of course, if $l^a\nabla_a\theta = 0$ is to be compatible with $\xi^a\nabla_a\theta = \chi^a\nabla_a\theta = 0$, then the congruence must share the same symmetries as the spacetime in that $\mathcal{L}_\xi l^a = \mathcal{L}_\chi l^a$. Unfortunately, even a spacetime satisfying all of our conditions (asymptotically flat, stationary, and axisymmetric) will admit many such congruences and thus many ways of extending r and θ into the interior. However, if we demand that the congruence be shear-free, thus restricting attention to algebraically special spacetimes, then (apart from the flat-spacetime case) the

Figure 12.2. An axisymmetric body allows a unique latitudinal angle $\theta(P)$ and a unique radial value $r(P)$ to be assigned to any point.

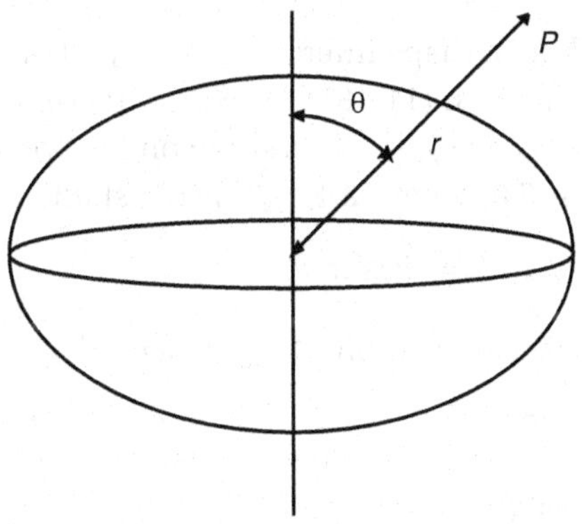

congruence will be unique: any other congruence will develop a shearing motion.

Putting all this together, we have shown that an asymptotically flat, stationary, axisymmetric, algebraically special spacetime possesses a unique latitudinal angle θ and a unique radial function r. Vacuum spacetimes with this long list of properties, together with *TP* invariance, do actually exist and, as we have already noted, are called Kerr spacetimes. As we shall see, they play a crucial role in the study of black holes. They also have a particularly simple structure in that they depend on only two numbers: a mass parameter m, and an angular momentum parameter a.

The angular momentum parameter a is defined by $a\sin^2\theta = -\hat{l}^a\hat{\chi}_a$ on $\mathcal{I}^+$. That a is actually constant on $\mathcal{I}^+$ can be shown to follow from the shear-free condition, $\hat{\sigma} = 0$. To gain some insight into its meaning, let us introduce a function u such that

$$\hat{\xi}^a\hat{\nabla}_a u = 1 \quad \text{and} \quad \hat{\chi}^a\hat{\nabla}_a u = \hat{l}^a\hat{\nabla}_a u = 0.$$

Under these conditions, u is determined up to $u \to u + \alpha(\theta)$, and we use this freedom to make $\hat{\nabla}^a u\hat{\nabla}_a\theta = 0$ on $\mathcal{I}^+$. Thus, on $\mathcal{I}^+$, we have

$$\hat{\nabla}^a u\hat{n}_a = 1 \quad \text{and} \quad \hat{\nabla}^a u\hat{\xi}_a = \hat{\nabla}^a u\hat{\chi}_a = \hat{\nabla}^a u\hat{\nabla}_a\theta = 0.$$

Furthermore, if $a = 0$, then

$$\hat{l}^a\hat{n}_a = 1 \quad \text{and} \quad \hat{l}^a\hat{\xi}_a = \hat{l}^a\hat{\chi}_a = \hat{l}^a\hat{\nabla}_a\theta = 0,$$

and hence, on comparing the above two sets of equations, we see that $\hat{l}_a = \hat{\nabla}_a u$. In other words, the congruence is surface-forming, and hence twist-free, if $a = 0$.

We therefore expect that a is related to $\hat{\rho}$ on $\mathcal{I}^+$. (We recall that, by construction, $\hat{\rho}$ is purely imaginary on $\mathcal{I}^+$ and is zero for a surface-forming congruence.) A calculation bears this out and gives (on $\mathcal{I}^+$) $\hat{\rho} = ia\cos\theta$. We already have shown that, for a shear-free congruence, $\rho = -1/(1+i\Sigma)$, where $D\Sigma = 0$ and $i\Sigma = \hat{\rho}$ on $\mathcal{I}^+$. We thus have

$$\rho = -\frac{1}{r + ia\cos\theta}. \tag{12.18}$$

It is perhaps interesting to note that this is the same as the corresponding equation (11.18) for a Kerr congruence in flat spacetime.

Since $\hat{\chi}_a\hat{l}^a = -a\sin^2\theta$ on $\mathcal{I}^+$, we have $\chi_a l^a \to -a\sin^2\theta$ as $r \to \infty$. But, by Theorem 12.1, $\chi_a l^a$ is constant along the congruence, and hence

$$\chi_a l^a = -a\sin^2\theta \tag{12.19}$$

everywhere. Similarly, since $\hat{\xi}_a\hat{l}^a = 1$ on $\mathcal{I}^+$,

$$\xi_a l^a = 1 \tag{12.20}$$

everywhere.

By TP invariance there must also exist another shear-free null vector field, l'_a, such that

$$l'^a \nabla_a r = 1, \qquad l'^a \nabla_a \theta = 0, \qquad l'^a \xi_a = -1, \qquad l'^a \chi_a = a \sin^2 \theta.$$

By expressing $\nabla_a r$ as a linear combination of l_a, l'_a, ξ_a, and χ_a we see that $\nabla_a r$ must be proportional to $l_a + l'_a$, and hence

$$\nabla_a r \nabla^a \theta = 0. \tag{12.21}$$

In order to compute quantities such as $\nabla_a \theta \nabla^a \theta$ in the interior we use the following useful theorem:

Theorem 12.4 *If functions f and g are constant along the congruence (i.e. $Df = Dg = 0$), then $h = \nabla_a f \nabla^a g$ is given by $h = h_0/\Sigma$, where $h_0 = \lim_{r\to\infty} h$ (limit taken along the congruence) and $\Sigma = r^2 + a^2 \cos^2 \theta$.*

Proof Let $(l^a, n^a, m^a, \bar{m}^a)$ be a null tetrad such that $Dm^a = Dn^a = 0$. Then, from the definition of ρ, we have

$$m^a \nabla_a l_b = -\bar{\rho} m_b + X l_b,$$

where X is some irrelevant function. Using this equation together with $Df = 0$ and $Dm^a = 0$, we have

$$\begin{aligned} D(m^a \nabla_a f) &= m^a D \nabla_a f = m^a l^b \nabla_b \nabla_a f = m^a l^b \nabla_a \nabla_b f \\ &= -m^a (\nabla_a l^b) \nabla_b f = \bar{\rho} m^b \nabla_b f. \end{aligned}$$

Writing

$$g^{ab} = l^a n^b + n^a l^b - m^a \bar{m}^b - \bar{m}^a m^b$$

and using $Df = Dg = 0$, we see that

$$h = g^{ab} \nabla_a f \nabla_b g = -m^a \nabla_f \bar{m}^b \nabla_g - \bar{m}^a \nabla_a f m^b \nabla_b h.$$

Thus

$$Dh = (\rho + \bar{\rho}) h.$$

Since $Dr = 1$ and $\rho = -(r + ia \cos \theta)^{-1}$, this implies

$$\frac{\partial h}{\partial r} = -\frac{2r}{r^2 + a^2 \cos^2 \theta} h$$

and hence

$$h = \frac{h_0}{r^2 + a^2 \cos^2 \theta}. \qquad \square$$

Since $\lim_{r\to\infty} \nabla_a\theta\nabla^a\theta = -1$ and $\lim_{r\to\infty} \nabla_a\theta\nabla^a u = 0$, this theorem gives

$$\nabla_a\theta\nabla^a\theta = -\Sigma^{-1} \tag{12.22}$$

and

$$\nabla_a\theta\nabla^a u = 0. \tag{12.23}$$

Let us now complete r, u, and θ to form a coordinate system by introducing a longitudinal function φ where

$$\xi^a\nabla_a\varphi = l^a\nabla_a\varphi = 0 \quad \text{and} \quad \chi^a\nabla_a\varphi = 1.$$

In terms of (u, r, θ, φ), the Killing vectors ξ^a and χ^a are given by

$$\xi^a = \partial_u^a \quad \text{and} \quad \chi^a = \partial_\varphi^a.$$

Since (u, θ, φ) forms a Bondi system on $\mathcal{I}^+$, it is easy to check that

$$\lim_{r\to\infty} \nabla_a\varphi\nabla^a\varphi = -\sin^{-2}\theta$$

and thus, by the above theorem,

$$\nabla_a\varphi\nabla^a\varphi = -\Sigma^{-1}\sin^{-2}\theta. \tag{12.24}$$

By proceeding in this way it is possible to find all contractions between $\nabla_a u$, $\nabla_a\theta$, and $\nabla_a\varphi$, and hence many of the components of the metric in terms of our coordinate system. A little thought shows that we now have enough information to calculate *all* components of the metric once we know $\nabla_a r\nabla^a r$. Using the vacuum equation, $R_{ab} = 0$, it is possible to show that

$$\nabla_a r\nabla^a r = -\frac{r^2 - 2mr + a^2}{\Sigma}, \tag{12.25}$$

where m is some constant known as the Kerr mass. Finally, a straightforward but tedious calculation gives

$$\begin{aligned} g = {} & \left(1 - \frac{2mr}{\Sigma}\right) du^2 + 2\,du\,dr + \frac{2mr}{\Sigma}(2a\sin^2\theta)\,du\,d\varphi \\ & - 2a\sin^2\theta\,dr\,d\varphi - \Sigma\,d\theta^2 \\ & - \left((r^2 + a^2)\sin^2\theta + \frac{2mr}{\Sigma}(a^2\sin^4\theta)\right) d\varphi^2, \end{aligned} \tag{12.26}$$

which is one form of the Kerr metric.

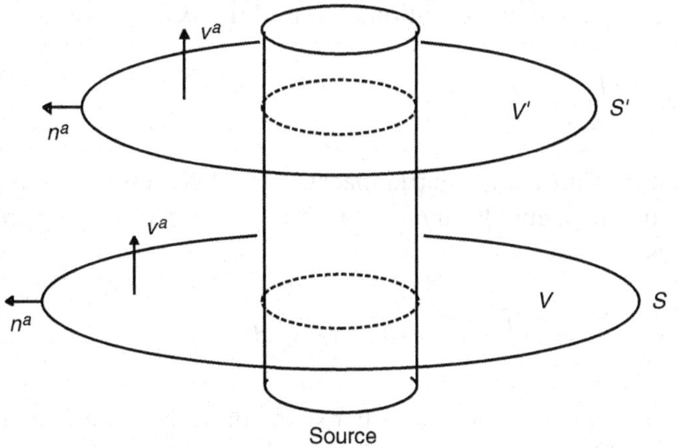

Figure 12.3. V and V' are two spacelike surfaces, with boundaries S and S', that pass through the source ($R_{ab} \neq 0$).

12.4 Energy and Intrinsic Angular Momentum

In Theorem 12.11 we showed that $\nabla^a f_{ab} = j_b$, where $j_b = R_{ab}\xi^a$ and $f_{ab} = \nabla_a \xi_b$. Therefore, if we take a situation like that depicted in Fig. 12.3, Gauss's theorem (Theorem 8.1) implies that j_a has a conserved charge given by

$$e_\xi = \int_V j_b v^a \, dV = \int_S \nabla_a \xi_b v^a n^b \, dS. \tag{12.27}$$

In other words, each Killing field ξ_a has an associated conserved charge e_ξ. This is a familiar situation in most branches of physics, where a symmetry is associated with a conserved quantity. For a timelike symmetry the conserved quantity is taken to be proportional to energy. Similarly, for a rotational symmetry the conserved quantity is taken to be proportional to angular momentum. In general relativity we therefore define the total energy of a stationary spacetime with Killing field ξ_a to be $E = ke_\xi$, where k is some constant. It remains only to determine this constant.

Writing Einstein's equation in the form

$$R_{ab} = -8\pi\left(T_{ab} - \tfrac{1}{2}Tg_{ab}\right),$$

we have

$$E = ke_\xi = -8\pi k \int_V (T_{ab} - \tfrac{1}{2}Tg_{ab})\xi^a v^b \, dV.$$

In a Newtonian approximation we can write $\xi^a = v^a$ and $T_{ab} = \rho v_a v_b$, and thus, by equation (8.38), we have

$$E = 2\int_V \left(T_{ab} - \tfrac{1}{2}Tg_{ab}\right)v^a v^b \, dV = \int_V \rho \, dV.$$

By comparing these two equations, we find that $k = -1/4\pi$ and hence

$$E = -\frac{1}{4\pi}\int_S \nabla_a \xi_b v^a n^b \, dS.$$

In the case of an axisymmetric spacetime with Killing field χ_a, we define the total intrinsic angular momentum by $J = k' e_\chi$ for some constant k'. This gives

$$J = k' e_\chi = -8\pi k' \int_V \left(T_{ab} - \tfrac{1}{2} T g_{ab}\right) \chi^a v^b \, dV.$$

To find the value of k' we go to a flat-spacetime approximation and compare this equation with equation (8.39):

$$J = -\int_V \left(T_{ab} - \tfrac{1}{2} T g_{ab}\right) \chi^a v^b \, dV = -\int_V T_{ab} \chi^a v^b \, dV.$$

This gives $k' = 1/8\pi$ and hence

$$J = \frac{1}{8\pi}\int_S \nabla_a \chi_b v^a n^b \, dS.$$

This leads to the following definition, first proposed by Komar (1959).

Definition 12.3 Given a spacetime with normalized Killing fields ξ^a (stationary) and χ^a (axisymmetric), then its total energy E and its total intrinsic angular momentum J are given by

$$E = -\frac{1}{4\pi}\int_S \nabla_a \xi_b v^a n^b \, dS, \tag{12.28}$$

$$J = \frac{1}{8\pi}\int_S \nabla_a \chi_b v^a n^b \, dS, \tag{12.29}$$

where S is a closed spacelike two-surface surrounding the source, and v^a (timelike and future-pointing) and n^a (spacelike and outward-pointing) are two orthogonal unit normals to S.

In the case of a Kerr spacetime, E and J can be calculated without too much difficulty by taking S to be a cross section of $\mathcal{I}^+$. This yields the interesting relation

$$J = aE, \tag{12.30}$$

which we shall find useful when we come to consider the properties of black holes.

EXERCISES

12.1 Use the results of Exercise 9.6(c) to prove equation (12.5).

12.2 Use equation (12.5) to show that $\hat{R} = 0$ for the conformal completion of flat spacetime with $\Omega = 1/r$.

12.3 Let $\hat{g}_{ab} = \Omega^2 g_{ab}$, where g_{ab} is a flat metric and $\Omega = (x_a x^a)^{-1}$. Show that $\hat{g}_{ab}$ is also a flat metric.

12.4 K_{ab} is said to be a Killing *tensor* if it is symmetric and satisfies $\nabla_{(a} K_{bc)} = 0$. If γ is a geodesic with affine tangent vector u^a, show that $K_{ab} u^a u^b$ is constant along γ.

12.5 Let $V^2 = \xi_a \xi^a$ where ξ^a is a Killing field. Show that

$$\nabla_c \nabla^c V^2 = -2 R_{ab} \xi^a \xi^b.$$

13

Schwarzschild Geometries and Spacetimes

An isolated physical body such as a star may be described in terms of a spacetime picture by an asymptotically flat spacetime, (M, g), containing a world tube representing the region of M occupied by the body. Inside the world tube, the energy–momentum tensor, T_{ab}, will be nonzero and should describe some physically reasonable matter distribution. Outside, in the vacuum, matter-free region, $T_{ab} = 0$, and hence, by Einstein's equation, $R_{ab} = 0$. The curvature, and hence the gravitational field, in the vacuum region is thus completely given by the Weyl tensor, that is, $R_{abcd} = C_{abcd}$.

While the situation described above serves as a framework for the discussion of isolated bodies in general relativity, it is too general to provide much real information; what we need is some simplification to make it more tractable to calculation. Fortunately, most stars are more or less spherically symmetric and stationary. We can thus obtain a useful, but considerably simpler, model by imposing spherical symmetry and stationarity. The problem of finding the most general spacetime satisfying these conditions was, in fact, essentially solved by Schwarzschild in 1916, not long after Einstein first proposed his general theory of relativity. In particular, he showed that the vacuum region outside a spherically symmetric and stationary body has a simple geometric structure, which we shall refer to as a **Schwarzschild geometry**, that depends on only one number, m, which may be identified with the total mass of the body. This result is similar to the Newtonian case, where the external gravitational field of *any* symmetric body is given simply by the inverse square law. It also soon became apparent that the vacuum region still possesses a Schwarzschild geometry even if the stationarity condition is dropped. This result, which allows a simple and tractable treatment of nonstationary (but spherically symmetric) processes such as gravitational collapse leading to a black hole, is known as **Birkhoff's theorem**. Again it has an analogue in both Newtonian gravity and electromagnetism. For example, an electrically charged bomb that explodes in a spherically symmetric fashion produces, in the vacuum region, an electric and gravitational field still satisfying the inverse square law.

In this chapter we shall derive Schwarzschild's result, *without imposing stationarity*, by means of the properties of null congruences discussed

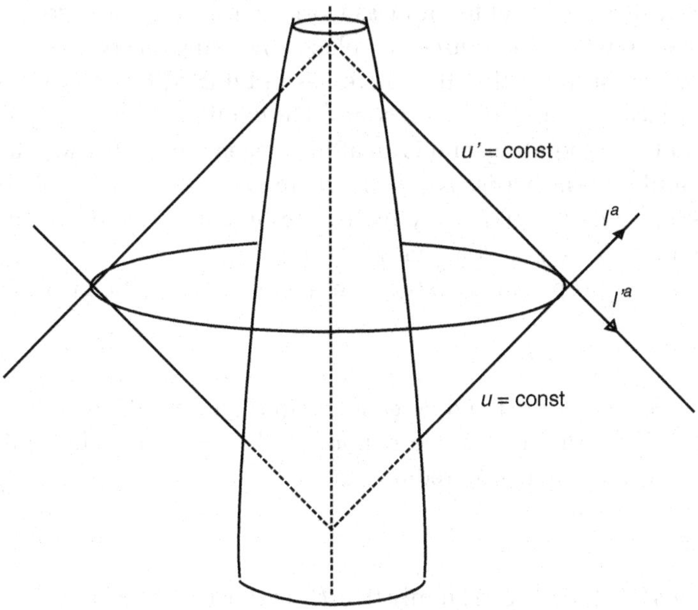

Figure 13.1. A future null cone ($u =$ const) and a past null cone ($v =$ const), with vertices on the center-of-mass world line, intersect in a two-sphere.

in Section 11.1. This has the advantage of providing concepts necessary for the discussion of gravitational collapse, and also provides a natural stepping stone to the Kerr spacetime, which we shall discuss later.

13.1 Schwarzschild Geometries

Consider a spherically symmetric, but not necessarily stationary, isolated gravitational source – for example, an exploding or imploding star. In the vacuum, (matter-free) region exterior to the source, the energy–momentum tensor vanishes and hence Einstein's equation implies $R_{ab} = 0$. In what follows we shall restrict attention to this region.

As Fig. 13.1 illustrates, future- and past-directed null cones with vertices on the center-of-mass world line of the source form two preferred families of null hypersurfaces, which, by spherical symmetry, intersect in a family of (metric) two-spheres. If a point p lies on such a two-sphere with area $A(p)$, we say that it has (positive) radius value $r(p)$ if $A(p) = 4\pi r(p)^2$. In this way we obtain a well-defined and unique radius function, r, on our spacetime.

Another way of defining such two-spheres, and hence the radius function, is to say that two points are contained in the same sphere if they are connected by a streamline of a spherical Killing vector. Though less physically intuitive, this definition has the advantage of not depending on the existence of a center-of-mass world line and hence carries over to the situation where the source is replaced by a singularity.

Let u be a null function that increases in future directions and whose level surfaces are the future null cones. The gradient, $l_a = \nabla_a u$, of u is thus null and future-pointing, and is an affine tangent vector to the generators of the null cones, that is, $Dl_a = 0$, where $D = l^a \nabla_a$. The function u is unique up to $u \to f(u)$ for any increasing function f, and this makes l_a unique up to $l_a \to \dot{f} l_a$. Note that $D\dot{f} = 0$.

Since r^2 is proportional to the area function A, equation (11.4) gives

$$Dr = -\rho r, \tag{13.1}$$

where $\rho = -\nabla_a l^a/2$ is the divergence of the future null cones [see equation (11.1)]. In the matter-free region, $R_{ab} l^a l^b = 0$, and the null Raychaudhuri equation (11.9), reduces to

$$D\rho = \rho^2. \tag{13.2}$$

Equations (13.1) and (13.2) imply that Dr is constant along the null generators and hence we can choose $\dot{f}$ such that $Dr = 1$. This makes l_a unique and implies

$$\rho = -\frac{1}{r} = -\tfrac{1}{2}\nabla_a l^a. \tag{13.3}$$

Similarly, we introduce a function u' that increases in *past* directions and whose level surfaces are *past* null cones, and write $l'_a = \nabla_a u'$ and $D' = l'^a \nabla_a$. Again, we see that $D' l'_a = 0$, that l'_a can be chosen such that $D' r = 1$, and that

$$\rho' = -\frac{1}{r} = -\tfrac{1}{2}\nabla_a l'^a. \tag{13.4}$$

We now have at our disposal two unique, geometrically defined null-vector fields, l^a (future-pointing) and l'^a (past-pointing), which satisfy

$$\begin{aligned} Dl_a &= D' l'_a = 0, \\ \nabla_a l^a &= \nabla_a l'^a = 2r^{-1}, \\ Dr &= D' r = 1. \end{aligned}$$

Note also that $\nabla_a l_b = \nabla_b l_a$ and $\nabla_a l'_b = \nabla_b l'_a$.

Let us now introduce a function U defined by

$$l_a l'^a = U^{-1}. \tag{13.5}$$

Since $Dl_a = 0$, we get $D(l_a l'^a) = l_a\, Dl'^a = -DU/U^2$. Furthermore, by expressing Dl'_a as a linear combination of l'_a and l_a (this is possible because of spherical symmetry) and using $D(l'_a l'^a) = 2l'_a Dl'^a = 0$, we get $Dl'_a = -(DU/U) l'_a$. Similarly, $D' l_a = -(D'U/U) l_a$.

From these relations we get

$$l^a l'^b \nabla_a \nabla_b U = l^a \nabla_a (l'^b \nabla_b U) - l^a \nabla_a l'^b \nabla_b U = D'DU - D'U\, DU$$

and, similarly, $l'^a l^b \nabla_a \nabla_b U = DD'U - DU\, D'U$. Thus, by torsionfreeness,

$$DD'U = D'DU.$$

By writing $\nabla_a r$, $\nabla_a U$, $\nabla_a l_b$, and $\nabla_a l'_b$ as appropriate linear combinations of l_a, l'_a, and g_{ab} (again, this is possible by spherical symmetry) and using the above relations, we get

$$\nabla_a r = U(l_a + l'_a), \tag{13.6}$$
$$\nabla_a U = -U(D'Ul_a + DUl'_a), \tag{13.7}$$
$$\nabla_a l_b = -D'Ul_a l_b - Ur^{-1}(l_a l'_b + l'_a l_b) + r^{-1} g_{ab}, \tag{13.8}$$
$$\nabla_a l'_b = -DUl'_a l'_b - Ur^{-1}(l_a l'_b + l'_a l_b) + r^{-1} g_{ab}. \tag{13.9}$$

The importance of these relations is that we can use them to calculate higher-order expressions such as $\nabla_c \nabla_b l_a$ in terms of D and D' derivatives of U. In particular, we can use them to express the curvature tensor in terms of $DD'U$, DU, and so on.

So far we have used only the fact that the components $R_{ab} l^a l^b$ and $R_{ab} l'^a l'^b$ of R_{ab} vanish in the matter-free region. In order to get an equation for the function U, we now use the fact that the cross term $R_{ab} l'^a l^b$ vanishes.

By the definition of the curvature tensor we have

$$\nabla_a \nabla_b l_c - \nabla_b \nabla_a l_c = R_{abcd} l^d,$$

which gives

$$l'^b (\nabla_a \nabla_b l^a - \nabla_b \nabla_a l^a) = R_{bd} l^d l'^b = 0.$$

Using this together with equations (13.6) to (13.9), we get

$$DD'U + 2\, D'Ur^{-1} = 0.$$

Similarly, by reversing the roles of l^a and l'^a, we also get

$$D'DU + 2DUr^{-1} = 0.$$

Since $DD'U = D'DU$, these two equations may be expressed as

$$DU = D'U \tag{13.10}$$

and

$$D'DU + 2DUr^{-1} = 0. \tag{13.11}$$

By spherical symmetry, U may be expressed as a function of r and u. Writing $U = f(u, r)$, we get

$$DU = \frac{\partial f}{\partial u} Du + \frac{\partial f}{\partial r} Dr = \frac{\partial f}{\partial r}$$

and

$$D'f = \frac{\partial f}{\partial u} D'u + \frac{\partial f}{\partial r} D'r = U^{-1} \frac{\partial f}{\partial u} + \frac{\partial f}{\partial r} r,$$

where we have used $Dr = D'r = 1$, $Du = l^a l_a = 0$, and $D'u = l'^a l_a = U^{-1}$. Equation (13.10) thus implies that f depends only on r. Similarly, equation (13.11) implies

$$\frac{d^2 f}{dr^2} + 2r^{-1} \frac{df}{dr} = 0,$$

which has a general solution of the form $U = f = a(1 - 2mr^{-1})$ where a and m are constants.

As we shall see shortly, m corresponds to the total gravitational mass of the source – but what about the constant a? For flat spacetime $a = -\frac{1}{2}$. This can be seen from the following argument: Let (t, x, y, z) be a flat-space coordinate system, and take the center-of-mass world line to be the t-axis. In this case $r = \sqrt{x^2 + y^2 + z^2}$, and, when pointing in the z-direction, l^a and l'^a will be given by

$$l^a = (1, 0, 0, 1) \quad \text{and} \quad l'^a = (-1, 0, 0, 1),$$

and hence $U^{-1} = l'_a l^a = -2$. Note also that $r_a r^a = -1$, where $r_a = \nabla_a r$. If, in the curved case, we demand that our spacetime be *asymptotically* flat for large r, then, by comparison with the flat case, we must have

$$\lim_{r \to \infty} U = -\tfrac{1}{2}, \tag{13.12}$$

and also

$$\lim_{r \to \infty} r_a r^a = -1. \tag{13.13}$$

Condition (13.12) gives $a = -\frac{1}{2}$, and hence for any physically reasonable spacetime corresponding to an *isolated* (and therefore asymptotically flat) spacetime we have

$$U = -\frac{1}{2}\left(1 - \frac{2m}{r}\right). \tag{13.14}$$

Furthermore, it can be shown that if U has this form, then all other components of R_{ab} vanish – which is fortunate for general relativity, for if this were not the case, then Einstein's equation in the form $R_{ab} = 0$ would admit no spherically symmetric, asymptotically flat solutions, and would therefore be devoid of any physical content. Indeed, R_{ab} vanishes if and only if $a = -\frac{1}{2}$, and hence Einstein's equation together with spherical symmetry actually implies asymptotic flatness.

Now that we have an expression for U in terms of r, equations (13.6) to (13.9) tell us everything about the geometry of a Schwarzschild spacetime. For example, writing $r_a = \nabla_a r$ and

$$\xi_a = -U(l_a - l'_a), \tag{13.15}$$

we find that

$$r_a r^a = -\left(1 - \frac{2m}{r}\right), \tag{13.16}$$

$$\xi_a \xi^a = \left(1 - \frac{2m}{r}\right), \tag{13.17}$$

$$\nabla_a \xi_b = \frac{m}{2Ur^2}(\xi_a r_b - \xi_b r_a). \tag{13.18}$$

Similarly, by taking higher derivatives, an expression for C_{abcd} can be obtained in terms of l_a and l'_a, or in terms of r_a and ξ_a.

Since the right-hand side of equation (13.18) is antisymmetric, $\nabla_{(a}\xi_{b)} = 0$, and hence ξ_a is a Killing field. Furthermore, by equation (13.17), ξ_a is timelike (at least for $r > 2m$) and is normalized in that $\xi_a \xi^a \to 1$ for $r \to \infty$. Thus, in accordance with Birkhoff's theorem, spherical symmetry implies stationarity. Also, equation (13.15) can be written as

$$\xi_a = -U(\nabla_a u - \nabla_a u') = -U\nabla_a t,$$

where

$$t = u - u', \tag{13.19}$$

and thus our spacetime is also static.

Let us now consider the meaning of the parameter m. As $r \to \infty$, we have $r_a r^a \to -1$ and $\xi_a \xi^a \to 1$, and hence r_a and ξ_a become unit vectors orthogonal to $r =$ const. Thus, by using Definition 12.3 for the energy and taking S to be an $r =$ const sphere where $r \to \infty$, equation (13.18) immediately gives $E = m$. Furthermore, as we shall now show, m is also equal to the total gravitational mass.

In Section 12.2 we showed that the acceleration of an inertial particle (for example, a freely falling apple) is given by

$$a_a = V^{-1} \nabla_a V,$$

where $V^2 = \xi_a \xi^a = 1 - 2m/r$ is the red-shift factor. For $r \to \infty$ this gives the inverse square law,

$$a_a = -\frac{m}{r^2} r_a, \tag{13.20}$$

thus showing that, at least asymptotically, m also is equal to the gravitational mass.

Finally, by introducing a coordinate system (u, r, θ, φ), where θ and φ are spherical coordinates such that

$$D\theta = D\varphi = D'\theta = D'\varphi = 0,$$

our orthogonallity relations imply that the metric has the form

$$g = \left(1 - \frac{2m}{r}\right) du^2 + 2\, du\, dr - r^2(d\theta^2 + \sin^2\theta\, d\varphi^2), \tag{13.21}$$

where $du\,dr$ is understood to be the symmetric tensor product of du and dr, that is, $du\,dr \leftrightarrow \nabla_{(a} u \nabla_{b)} r$.

13.2 Geodesics in a Schwarzschild Spacetime

In order to predict the motion of light rays and inertial particles in the external gravitational field of a symmetric body, we need to solve the geodesic equation for null and timelike geodesics. This can be done by first introducing a suitable coordinate system and then calculating the Christoffel symbols from the metric. However, considering the high degree to symmetry possessed by a Schwarzschild spacetime, this would be a foolish thing to do – unless, of course, one enjoys tedious calculations. A better and much simpler method is to use Theorem 12.1 together with ξ^a and an axisymmetric Killing field χ^a. According to this theorem, $E = w^a \xi_a$ and $J = w^a \chi_a$ are constant along a geodesic γ, where w^a is its four-velocity vector. (In the case of a null geodesic, we take w^a to be an affine tangent vector to γ.) However, by spherical symmetry, we can choose χ^a in many different ways.

Given a point P on γ, we can restrict this choice by taking χ^a to be *equatorial* in that $\chi_a \chi^a = -r^2$ (see Fig. 13.2). This leaves us with a circle's worth of such Killing fields, but only one of them, χ_P^a say, will satisfy

$$w^a = a\xi^a + br^a + c\chi^a \tag{13.22}$$

at P. By taking some other point, Q, on γ, we obtain in the same

Figure 13.2. An equatorial and a nonequatorial Killing vector on a sphere of radius r.

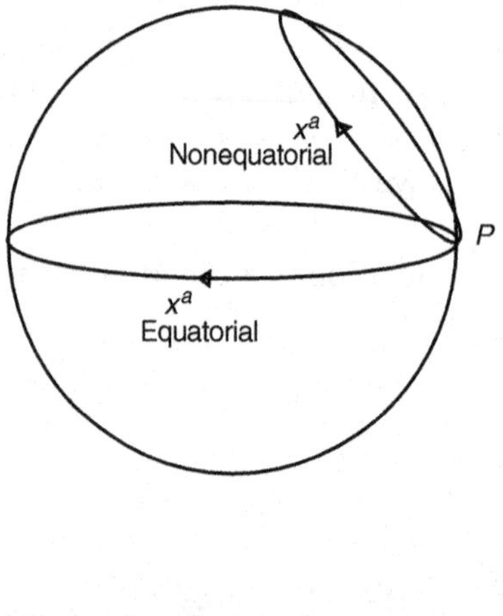

way another equatorial Killing field χ_Q^a. But, by the symmetry of the situation, we see that χ_Q^a must equal χ_P^a at P, and hence χ_P^a is independent of the choice of P on γ. In other words, *a geodesic determines a unique equatorial Killing field* χ^a. This corresponds to the fact that in Newtonian theory an orbit lies in a plane.

Since $w_a w^a = 1, \xi_a \xi^a = 1 - 2m/r$, and $\chi_a \chi^a = -r^2$, equation (13.22) gives

$$w_a w^a = \left(1 - \frac{2m}{r}\right)(a^2 - b^2) - c^2 r^2 = 1. \tag{13.23}$$

Furthermore, since ξ^a and χ^a are Killing vectors, Theorem 12.1 implies that

$$E = w^a \xi_a = a\left(1 - \frac{2m}{r}\right)$$

and

$$J = w^a \chi_a = cr^2$$

are constant along the geodesic.

Let us now introduce a longitudinal angle, φ, such that $\chi^a \nabla_a \varphi = 1$. The rate of change of r and φ (with respect to proper time) along the geodesic is given by

$$\dot{r} = w^a \nabla_a r = w^a r_a = -b\left(1 - \frac{2m}{r}\right)$$

and

$$\dot{\varphi} = w^a \nabla_a \varphi = c.$$

By eliminating c and b we obtain

$$\dot{r}^2 = E^2 - 1 + 2\frac{m}{r} - \frac{J^2}{r^2} + 2\frac{mJ^2}{r^3} \tag{13.24}$$

and

$$J = \dot{\varphi} r^2. \tag{13.25}$$

Following the device commonly used in Newtonian theory, we now introduce a new variable $\omega = r^{-1}$ and use equation (13.25) to replace proper time by φ as a parameter. This gives

$$\left(\frac{d\omega}{d\varphi}\right)^2 = \frac{1}{J^2}(E^2 - 1 + 2m\omega - J^2\omega^2 + 2mJ^2\omega^3), \tag{13.26}$$

which is the required form of the geodesic equation in a Schwarzschild spacetime.

For the orbit of the earth around the sun we know the values of m (the mass of the sun), r (the radius of the earth's orbit), and $\dot{\varphi}$ (the angular velocity of the earth around the sun). By converting these values to natural units where $c = G = 1$ and using equation (13.25), we find $2m\omega \approx 10^{-8}$, $J^2\omega^2 \approx 10^{-8}$, and $2mJ^2\omega^3 \approx 10^{-16}$. Thus, in the case of the earth – and, in fact, any planet in the solar system – the term $2mJ^2\omega^3$ may be considered to be small compared with the other terms on the right-hand side of equation (13.24). By neglecting this term and writing $h = E^2 - 1$ we get

$$\left(\frac{d\omega}{d\varphi}\right)^2 = \frac{1}{J^2}(h + 2m\omega - J^2\omega^2), \tag{13.27}$$

which may be recognized as the Newtonian equation of the orbit of a particle in spherically symmetric gravitational field, thus confirming that general relativity is consistent with Newtonian gravity to a very high degree of accuracy. The solution of this equation is given by

$$\omega = \frac{m}{J^2}(1 + e\cos\varphi), \tag{13.28}$$

where

$$e^2 = 1 + \frac{hJ^2}{m^2} \tag{13.29}$$

and the arbitrary constant has been chosen so that $d\omega/d\varphi = 0$ at $\varphi = 0$. If $h < 0$, this gives an ellipse (see Fig. 13.3). Note that

$$L = (\text{major axis}) \times \frac{1 - e^2}{2} = \frac{J^2}{m}. \tag{13.30}$$

The same method gives the equation of motion of a *null* geodesic, the only difference being that the left-hand side of equation (13.23) is replaced by zero. The required equation is given by

$$\left(\frac{d\omega}{d\varphi}\right)^2 = \frac{1}{l^2} - \omega^2 + 2m\omega^2, \tag{13.31}$$

where $l = J/E$. Again the Newtonian equation is found by omitting the last term, and the Newtonian solution is just

$$\omega = \frac{1}{l}\sin\varphi, \tag{13.32}$$

where the arbitrary constant has been chosen so that $d\omega/d\varphi = 0$ at $\varphi = \pi/2$. This is, of course, simply the equation of a straight line – in the Newtonian approximation

Figure 13.3. The Newtonian orbit of a planet. Perihelion occurs at point A ($\varphi = \pi$).

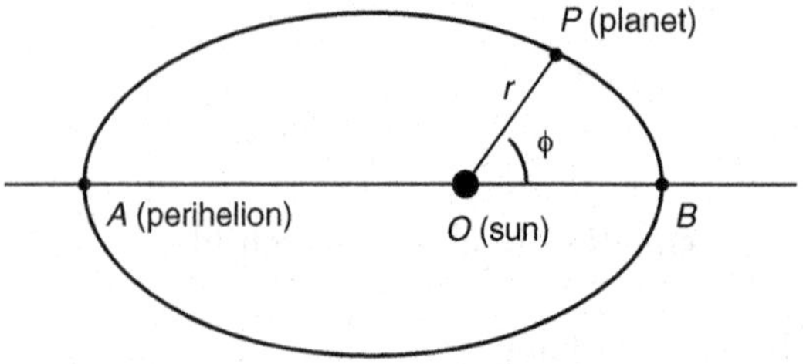

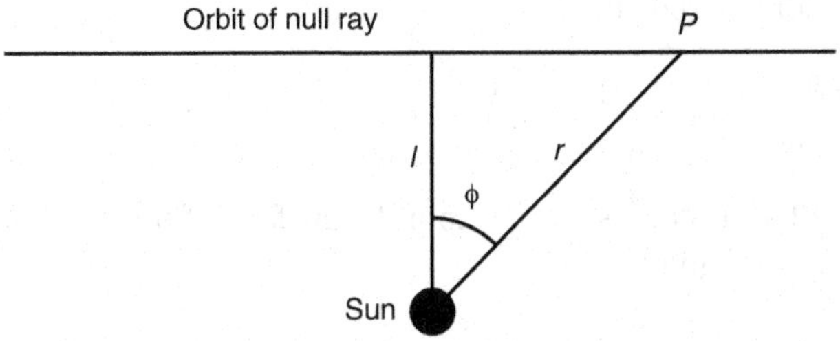

Figure 13.4. Orbit of a null ray. Least distance from Sun is $l = J/E$.

light still travels along a straight line. Note that l gives the shortest distance between the null ray and the sun (see Fig. 13.4).

13.3 Three Classical Tests of General Relativity

As we have seen, the theory of general relativity is in agreement with Newtonian gravity to high degree of accuracy. It is also intellectually a very satisfying theory in that it accommodates gravity in a spacetime picture and therefore resolves difficulties concerning the conflict between the notions of absolute space and time and the invariance of the speed of light. However, in order for a new theory to be acceptable it must stick its neck out and make *new* and *testable* predictions. In this section we shall consider three such predictions of general relativity – namely, gravitational red shifts, perihelion precession, and the bending of light rays – all of which have been experimentally verified to a high degree of precision.

We have already considered gravitational red shifts in a qualitative way in Section 12.2. There we showed that a photon suffers a red shift as it moves away from a gravitating body. In particular, we showed that if a photon is emitted with a frequency ω_0 from a point p_0 and received with a frequency ω_1 at a point p_1 then

$$\frac{\omega_1}{\omega_0} = \frac{V(p_0)}{V(p_1)},$$

where $V^2 = \xi^a \xi_a$ is the red shift factor. For a Schwarzschild spacetime,

$$V^2 = \xi^a \xi_a = 1 - \frac{2m}{r} \tag{13.33}$$

and hence

$$\frac{\omega_1}{\omega_0} = \left(1 - \frac{2m}{r_0}\right)^{-1/2} \left(1 - \frac{2m}{r}\right)^{-1/2}.$$

If r_0 and r_1 are small compared to $2m$, this gives the following approximate expression for the red shift:

$$\frac{\Delta\omega}{\omega} = \frac{\omega^0 - \omega_1}{\omega_0} = m\left(\frac{1}{r_0} - \frac{1}{r_1}\right). \tag{13.34}$$

In the case of a photon escaping from the surface of the sun and received on the earth, this gives

$$\frac{\Delta\omega}{\omega} = 2.12 \times 10^{-6}.$$

Observations of the sun's spectrum give a red shift of this order, but due to random fluctuations there is some difficulty in interpreting these results. Similar remarks hold for white dwarfs, which, because of their small radius compared with their mass, have a more pronounced red shift. However, recent terrestrial experiments using the Mössbauer effect have confirmed the gravitational red shift to very high degree of accuracy.

Let us now turn to perihelion precession. As we showed in the previous section, the equation of motion of a orbit in a Schwarzschild spacetime is given by

$$\left(\frac{d\omega}{d\varphi}\right)^2 = \frac{1}{J^2}(E^2 - 1 + 2m\omega - J^2\omega^2 + 2mJ^2\omega^3). \tag{13.35}$$

This reduces to the Newtonian equation if the term $2mJ^2\omega^3$ is omitted. Let us see what happens, at least to first order, if we retain this term. Since

$$\omega = \frac{m}{J^2}(1 + e\cos\varphi) \tag{13.36}$$

is the solution in the Newtonian case, we seek a solution of the form

$$\omega = \frac{m}{J^2}(1 + e\cos\varphi) + U \tag{13.37}$$

where U is a first-order perturbation. Substituting this into equation (13.35) and retaining only first-order terms in u, we obtain

$$U = \frac{m^3}{J^4}\left[\frac{1+3e^2}{e}\cos\varphi + 3\left(1 + \frac{e^2}{2}\right) - \frac{e^2}{2}\cos 2\varphi + 3e\varphi\sin\varphi\right]. \tag{13.38}$$

The interesting term in this expression is $3e\varphi\sin\varphi$, which is nonperiodic and thus produces a *cumulative* effect; the other terms are periodic and thus produce only small, noncumulative wobbles in the orbit. For the Newtonian orbit, perihelion is reached at $\varphi = \pi$, where $d\omega/d\varphi = 0$. However, for the relativistic orbit we have

$$\frac{d\omega}{d\varphi} = -\frac{me}{J^2}\sin\varphi + \frac{m^3}{J^4}\left(-\frac{1}{e}\sin\varphi + e^2\sin 2\varphi + 3e\varphi\cos\varphi\right),$$

which is nonzero at $\varphi = \pi$ but zero at a slightly advanced angle $\varphi = \pi + \varepsilon$. To find ε we solve $d\omega/d\varphi = 0$ to first order in ε and get

$$\varepsilon = \frac{3m^2}{J^2}\pi = \frac{3m}{L}\pi,$$

where L is given by equation (13.30). Thus, in one whole orbit, perihelion advances by

$$2\varepsilon = \frac{6m\pi}{L} \text{ radians.} \qquad (13.39)$$

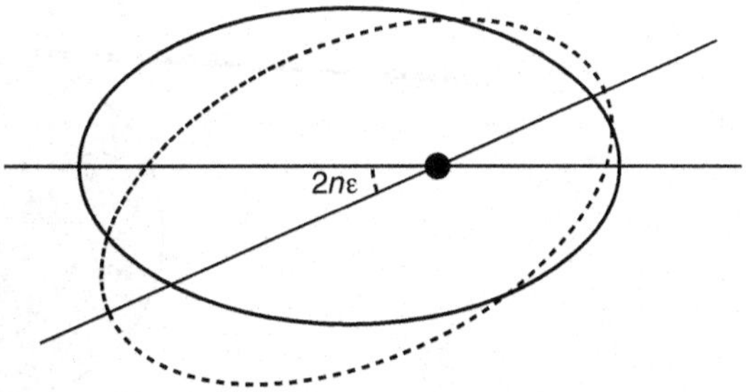

Figure 13.5. Perihelion advance. The solid curve represents the initial orbit, and the dashed curve the orbit after n periods.

Since this effect is cumulative, the advance in perihelion after n orbits is given by

$$2n\varepsilon = n\frac{6m\pi}{L} \text{ radians.} \qquad (13.40)$$

(See Fig. 13.5). This is the celebrated **perihelion advance** predicted by general relativity. The advance is largest when L is as small as possible, which is the case for the planet Mercury. In this case equation (13.40) gives an advance in perihelion of 43 seconds of arc per century, which is, up to experimental error, equal to the observed value, which had been known since the nineteenth century and originally considered an anomaly, since it could not be explained by Newtonian gravity.

Let us now turn to the bending of light rays. As we showed in the previous section, the orbit in this case is given by

$$\left(\frac{d\omega}{d\varphi}\right)^2 = \frac{1}{l^2} - \omega^2 + 2m\omega^2, \qquad (13.41)$$

where l may be taken to be its minimum distance from the center of the sun. Neglecting the term $2m\omega^2$, the solution is

$$\omega = \frac{1}{l}\sin\varphi. \qquad (13.42)$$

We thus look for a solution of the full equation of the form

$$\omega = \frac{1}{l}\sin\varphi + U. \qquad (13.43)$$

Substituting this into the full equation and retaining only first-order terms, we get

$$U = \frac{3m}{2l^2}\left(1 + \tfrac{1}{3}\cos 2\varphi\right). \qquad (13.44)$$

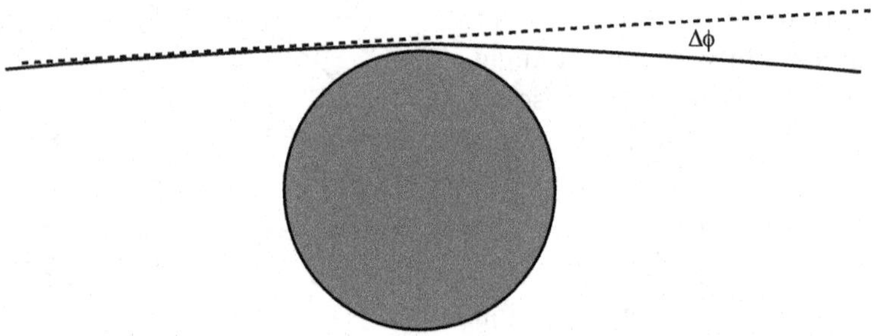

Figure 13.6. A light ray is deflected by an angle $\Delta\varphi$ as is grazes the surface of the sun.

The effect of this term is to produce a deflection in the straight line given by Newtonian equation (13.42) (see Fig. 13.6). To calculate this deflection we find the limit, φ_∞, of φ as $r \to \infty$ (i.e. $\omega \to 0$). Since φ_∞ is first-order (it is zero in the Newtonian case) we may write $\sin\varphi_\infty = \varphi_\infty$ and $\cos 2\varphi_\infty = 1$, and equations (13.43) and (13.44) give

$$0 = \frac{\varphi_\infty}{l} + \frac{3ml^2}{2}\left(1 + \tfrac{1}{3}\right),$$

that is,

$$\varphi_\infty = -\frac{2m}{l}.$$

Since the total deflection, $\Delta\varphi$, is *twice* φ_∞, we have

$$\Delta\varphi = \frac{4m}{l}. \tag{13.45}$$

With m equal to the mass of the sun and l equal to its radius, this gives

$$\Delta\varphi = 1.75',$$

which is in agreement with an observation first performed during an eclipse of the sun in 1919 to about 1%.

13.4 Schwarzschild Spacetimes

As we have seen, the matter-free region outside any spherically symmetric source has a particularly simple structure, which we called a Schwarzschild geometry. The spacetime structure in the interior will, of course, depend on the nature of the source and may be extremely complicated, but the exterior region will always possess a Schwarzschild geometry. For example, the exterior region of a large dilute body and the exterior region of a small dense body, both with the same mass m, will both possess the same Schwarzschild geometry of mass m – the only difference being that the former region will be smaller than the latter, i.e. the latter region will be an *extension* of the former. Thus, if we have a spacetime M (inextendible by definition) whose geometry is Schwarzschild *everywhere*, then the exterior region of *any* spherically symmetric source will correspond to some asymptotically flat subspace of M. We refer to M as a **Schwarzschild spacetime**. In the literature it is sometimes called a maximally extended Schwarzschild or a Kruskal spacetime. A Schwarzschild spacetime thus provides a universal, and very useful, setting for the discussion of any isolated spherically symmetric process within general relativity – especially gravitational collapse and black holes.

A Schwarzschild spacetime M is thus spherically symmetric, and its Ricci tensor R_{ab} vanishes everywhere. In Section 13.1 we proved that M has the following properties:

(i) There exist two preferred, surface-forming null congruences corresponding to outgoing and incoming radial null rays. We shall refer to these null rays as **principal null directions.**

(ii) There exists a preferred radial function, r, which serves as an affine parameter along *both* principal null directions, and whose gradient $r_a = \nabla_a r$ satisfies

$$r_a r^a = -\left(1 - \frac{2m}{r}\right). \tag{13.46}$$

Thus r^a is spacelike for $r > 2m$, null or zero for $r = 2m$, and timelike for $r < 2m$. The radial value $r_S = 2m$ is called the **Schwarzschild radius**, and the three-surface H given by $r = r_S$ is called the **Schwarzschild horizon**. Since $r_a r^a = 0$ on H, H is a null surface.

(iii) There exists a static Killing vector field ξ^a such that

$$\xi_a \xi^a = 1 - \frac{2m}{r} \tag{13.47}$$

and $\xi^a r_a = \xi^a \nabla_a r = 0$. Thus ξ^a is timelike for $r > 2m$, null or zero for $r = 2m$, and spacelike for $r < 2m$.

(iv) Finally, a calculation shows that

$$R_{abcd} R^{abcd} = \frac{48m^2}{r^6}. \tag{13.48}$$

Since, by our definition of a spacetime, the metric and hence the curvature tensor is well defined at all events, there exists no point $p \in M$ such that $r(p) = 0$; r is thus strictly positive. We can, however, add points to M, *considered only as a manifold*, on which $r = 0$. The set of such points is called **the Schwarzschild singularity**.

Using r^a and ξ^a, we now distinguish the following regions of M: Region I where $r > 2m$ and ξ^a is future-pointing, region I′ where $r > 2m$ but

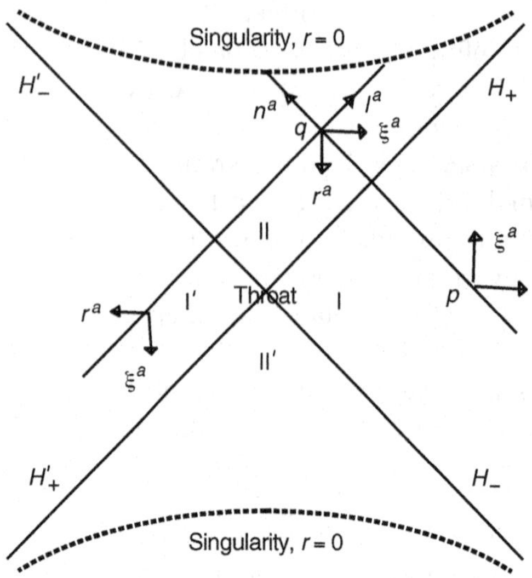

Figure 13.7. The Kruskal diagram for a Schwarzschild spacetime. Each point represents a two-sphere, and lines at 45° represent radial null geodesics.

where ξ^a is past-pointing, region II where $r < 2m$ and r^a is past-pointing, and finally region II′ where $r < 2m$ but where r^a is future-pointing. Notice that II and II′ contain two distinct singular regions $r \to 0$.

We illustrate these regions by means of a **Kruskal diagram** shown in Fig. 13.7. In such a diagram a point represents a two-sphere and lines at 45° represent radial null rays, that is, principal null directions. This, in particular, highlights the causal structure of the spacetime. The boundaries between the various regions are denoted by H_+, H_-, H'_+, and H'_-. These boundaries meet at a two-sphere, called the **Schwarzschild throat**, given by $r = 2m$ and $r^a = 0$.

In order to show that all the above regions exist, and also to illustrate some of their properties, let us make a tour of M along radial null rays. Starting from a point p (see Fig. 13.7) in I, we travel radially inwards (decreasing r) in a future direction along a null ray γ, that is, we travel in the direction of a tangent vector n^a such that $n^a \nabla_a r = n^a r_a < 0$ and $n^a \xi_a > 0$ at p. Since r serves as an affine parameter for γ, we may choose n^a to be affine and such that $n^a r_a = -1$, on the whole of γ. In this case, Theorem 12.1 implies that $n^a \xi_a$ is constant on γ and thus positive on the whole of γ. After crossing $r = 2m$, r^a becomes timelike, and since $n^a r_a = -1$, it must be past-pointing. We have thus crossed H_+ and entered region II. The null ray will now fall into the future component of the singularity. Note that, even though γ is *inextendible* in the future, its affine parameter (r in this case) is bounded in the future. This is simply because there is no spacetime event for which $r = 0$.

Suppose now that, at a point q in II, γ intersects another radial null ray, γ', traveling in the opposite direction with future-pointing (affine) tangent vector l^a. Since r^a is past-pointing in II, we must have $l^a r_a = l^a \nabla_a r < 0$, and thus r also decreases in direction l^a. From the point of view of flat spacetime, this represent a very odd situation. It means that the area function A decreases not only in an ingoing future null direction (as one might expect), but also in the opposite outgoing future null direction (which is strange). In terms of the divergence functions, ρ of l^a and ρ' of n^a, this means that $\rho > 0$ (l^a contracting) and $\rho' > 0$ (n^a also contracting). In a "normal" situation, in flat space for example, we have of course $\rho < 0$ and

$\rho' > 0$. In general, a closed spacelike two-surface whose two divergence functions corresponding to its two normal, future-pointing null vectors are both positive is called a **trapped surface**. All two-spheres in region II (i.e., points on the Kruskal diagram) are thus trapped surfaces.

Let us now move down the null ray γ' into the past (increasing r). Since r also serves as an affine parameter on γ', we may assume that l^a is affine and that $l^a r_a = -1$ on the whole of γ'. Furthermore, since γ' travels in the opposite direction to γ and $n^a \xi_a > 0$ at q, we have $l^a \xi < 0$ at q. Again by Theorem 12.1, $l^a \xi_a$ will be constant on γ' and thus negative on the whole of γ'. As we move down γ', r will increase, and after crossing $r = 2m$, ξ^a will become timelike. Since $l^a \xi < 0$, it will also be past-pointing. We have thus crossed H'_- and entered region I$'$.

Another way of getting between regions I and I$'$ is along a radial, *space-like* geodesic γ, starting at a point p in I, that is, along a streamline of r^a starting from p. That such a curve is actually a geodesic can be seen as follows:

$$\begin{aligned} r^b \nabla_b r_a &= r^b \nabla_b \nabla_a r = r^b \nabla_a \nabla_b r = r^b \nabla_a r_b \\ &= \tfrac{1}{2} \nabla_a (r^b r_b) = -\tfrac{1}{2} \nabla_a \left(1 - \frac{2m}{r}\right) = -\frac{m}{r^2} r_a. \end{aligned} \tag{13.49}$$

The radial curve γ is thus a geodesic, but r^a is not an *affine* tangent vector to γ. Furthermore, since ξ_a is proportional to $\nabla_a t$, where $t = u - u'$ [see equation (13.19)], and $\xi^a \nabla_a r = \xi^a r_a = 0$, r^a must satisfy $r^a \nabla_a t = 0$. In other words, the vector field r^a, and hence its streamlines, lie in the level surfaces of the time function t.

In order to make our journey from I to I$'$, we first introduce an *affine* length parameter s along γ and a corresponding unit tangent vector s^a, satisfying $s^a \nabla_a s = 1$ and $s^a s_a = -1$. Thus

$$s^a = \pm \left(1 - \frac{2m}{r}\right)^{-1/2} r^a. \tag{13.50}$$

The rate of change of r with respect to s along γ is given by

$$\dot{r} = s^a \nabla_a r = s^a r_a = \pm \left(1 - \frac{2m}{r}\right)^{1/2},$$

and hence

$$\dot{r}^2 = 1 - \frac{2m}{r}. \tag{13.51}$$

Starting at p in I with s^a inward-pointing ($\dot{r} < 0$) and moving along γ, we see that r decreases until we reach $r = 2m$, where $\dot{r} = 0$, and hence $r^a = 0$, that is, we have reached the Schwarzschild throat. After this point r begins to increase ($\dot{r} > 0$) and we enter region I$'$.

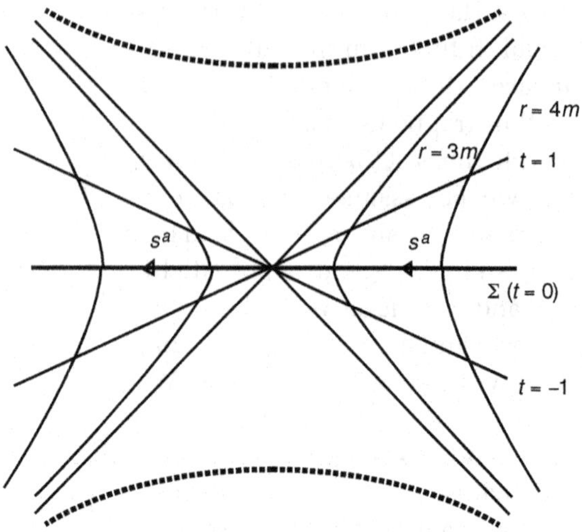

Figure 13.8. The spacelike three-surface Σ passes through the Schwarzschild throat.

Since s^a is spacelike along the whole of γ, and γ lies in some level surface, Σ say, of the time function t (for all directions of γ), we see that Σ is a spacelike three-surface that passes through the Schwarzschild throat (see Fig. 13.8). As we move along γ, we pass through a sequence of concentric two-spheres contained in Σ whose radius function initially decreases until we reach $r = 2m$, after which it starts to increase. We thus have a topological situation like that illustrated in Fig. 13.9.

The importance of the three-surface Σ lies in the fact that any causal (timelike or null) curve through any point p of M always intersects Σ if extended far enough. Thus, if p lies to the future of Σ, then all causal chains from p invariably lead back to Σ. This means that the conditions at p can be predicted if we know the full set of initial conditions on Σ. In general, a noncompact, spacelike three-surface with this property is called a **global Cauchy hypersurface**, and a spacetime containing such a surface is said to be **predictable**. A Schwarzschild spacetime thus has the highly desirable property of being predictable in this sense–as is, of course, flat spacetime.

EXERCISES

13.1 Express the Schwarzschild metric in terms of the coordinate systems (u', r, θ, φ), (t, r, θ, φ), and (u, u', θ, φ)

13.2 Let γ be a radial ray tangent to l'^a. Show that the restriction of u to γ has the form

$$u = -2r - 4m \times \ln|r - 2m| + C,$$

where C is some constant.

13.3 The previous exercise shows that u is not well defined on H_+ and H'_+ in that $u \to \infty$ along γ. Show that there exists a null coordinate system (w, w', θ, φ) which is well defined everywhere except on T.

Figure 13.9. As we move along a radial geodesic in Σ, the radius function initially decreases until we reach $r = 2m$. After this point we enter region I′ and the radius function begins to increase.

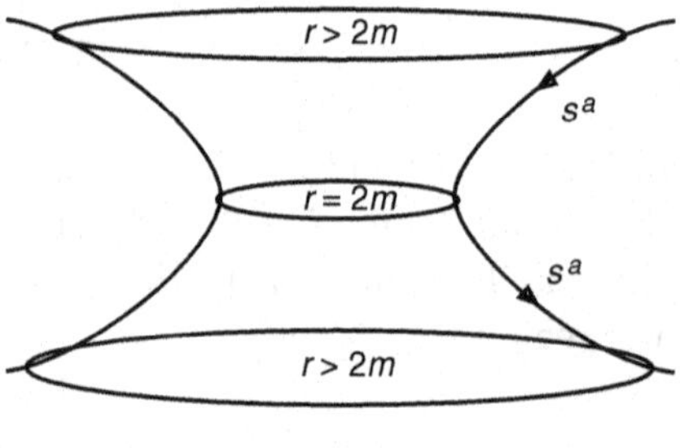

13.4 Show that closed orbits cannot exist in regions II and II$'$.

13.5 Let w^a be the four-velocity of a timelike, radial orbit. Show that

$$\dot{r}^2 = E^2 - 1 + \frac{2m}{r},$$

where $\dot{r} = w^a \nabla_a r$ and $E = w^a \xi_a$.

13.6 Starting from rest at infinity, an observer O falls radially toward a spherically symmetric star. When he reaches a radius value r_0 he receives a radial photon, which was transmitted with frequency ω_0 by a stationary observer O_∞ at infinity. What is the photon's frequency according to O? If O transmits a photon of frequency ω_0 radially outwards, what is the photon's frequency when received by O_∞?

13.7 Show that the escape velocity v_0 from the surface of a spherical star of radius r_0 and mass m is given by $v_0^2 = 2m/r_0$. Note that this is the same as the corresponding Newtonian expression.

13.8 Show that

$$\nabla_a l_a = \frac{m}{r^2} l_a l_b + \left(1 - \frac{2m}{r}\right) l_{(a} l'_{b)} + \frac{1}{r} g_{ab}, \tag{E13.1}$$

$$\nabla_a l'_a = \frac{m}{r^2} l'_a l'_b + \left(1 - \frac{2m}{r}\right) l_{(a} l'_{b)} + \frac{1}{r} g_{ab}, \tag{E13.2}$$

$$2r_a = \left(1 - \frac{2m}{r}\right)(l_a + l'_a). \tag{E13.3}$$

14

Black Holes and Singularities

14.1 Spherical Gravitational Collapse

One of the most remarkable predictions of general relativity is that, under extreme – but by no means unobtainable – conditions, gravitational collapse can lead inexorably to arbitrarily high densities and, ultimately, to a spacetime singularity. Gravitational collapse takes place when the internal pressure in a body (e.g., a star) is insufficient to counteract the inward pull of gravity. Since the pressure increases as the body contracts, one might expect that there will always come a point when this will be sufficient to prevent further contraction and that the star will settle down in some stable but denser state. This is indeed the case for a star of mass equal to that of the sun. The theory of stellar evolution tells us that such stars can reach a final equilibrium state as a white dwarf or a neutron star. However, for slightly larger stars, no such final equilibrium state is possible, and in such a case the star will contract beyond a certain critical point – the point of no return – where complete gravitational collapse leading to a spacetime singularity is inevitable.

In this section we restrict attention to the idealized case of spherically symmetric collapse, but, as we shall see later, the same phenomenon also occurs in a more general setting.

In order to gain an intuitive view of gravitational collapse, we first consider the case of Newtonian gravity. If a spherical body has mass m and radius r_0, then the escape velocity v_0 from its surface is given by

$$v_0^2 = \frac{2m}{r_0}.$$

In the case of the sun, v_0 is well below the speed of light, but, as Laplace first pointed out in 1798, a body of the same density of the sun but 250 times its radius would have an escape velocity equal to the speed of light, that is, $1 = 2m/r_0$, or $r_0 = 2m$. This is the Schwarzschild radius, which we discussed previously. The gravitational attraction of such a body would be so great as to prevent any material particle, including light, from escaping to infinity – we have, effectively, a Newtonian black hole. This is about as much as one can say within the context of Newtonian gravity and corresponding notions of absolute space and time, but if we consider the same situation within a spacetime context we come to a remarkable conclusion. A fundamental assumption, which forms one of the cornerstones of the spacetime picture of the world, is that any observer always finds himself at the center of the wavefront of an emitted flash of light – in other words, all emitted photons diverge away from him. In particular, this will be the case for an observer on a collapsing star whose radius has just fallen below the Schwarzschild radius. A photon emitted in an outward radial direction will essentially remain suspended at the Schwarzschild radius, but, to the observer, will appear to be moving away from him. He would therefore, inevitably, have an inward motion, and the star will continue to contract with ever increasing acceleration.

Being based on incompatible bedfellows in the form of Newtonian and relativistic physics, this argument should, of course, be taken with a large grain of salt. However, as we shall now show, the essential conclusions remain the same within an entirely relativistic, curved-spacetime picture.

We first note that the radius r_0 of the star and its mass m are still well defined in a curved-spacetime setting. The radius is defined by $A = 4\pi r_0^2$, where A is the surface area of the star, and the mass is well defined because the exterior, matter-free region is described by a Schwarzschild spacetime – we simply take m to be the Schwarzschild mass. Let $r_0(\tau)$ be the radius of the star at proper time τ according to a particle on the star's surface, and let w^a be the particle's four-velocity. Thus $\dot{r}_0 = w^a \nabla_a r_0$ gives the rate of increase of the star's radius according to a hypothetical (and unfortunate) observer sitting on the star's surface. Let us further assume that, as in Fig. 14.1, r_0 has dropped below the Schwarzschild radius $2m$.

In the region $r < 2m$, but outside the star, we have $\xi^a \xi_a = 1 - 2m/r < 0$ and $r^a r_a = -(1 - 2m/r) > 0$, which imply that ξ^a is spacelike and r^a is timelike. Since any (radial) four-velocity vector can be written as $w^a = a\xi^a + br^a$, the equations $w^a w_a = 1$ and $\xi^a r_a = 0$ imply

$$b^2 = a^2 - \left(1 - \frac{2m}{r}\right)^{-1} > 0$$

and

$$\dot{r} = w^a r_a = w^a \nabla_a r$$

$$= -b\left(1 - \frac{2m}{r}\right) \neq 0.$$

Taking w^a to be the four-velocity of a particle on the star's surface, we see that $\dot{r}_0 \neq 0$ and hence any

Figure 14.1. If the radius of a star falls below $2m$, it will inevitably collapse to a singularity within a finite time according to an observer on its surface.

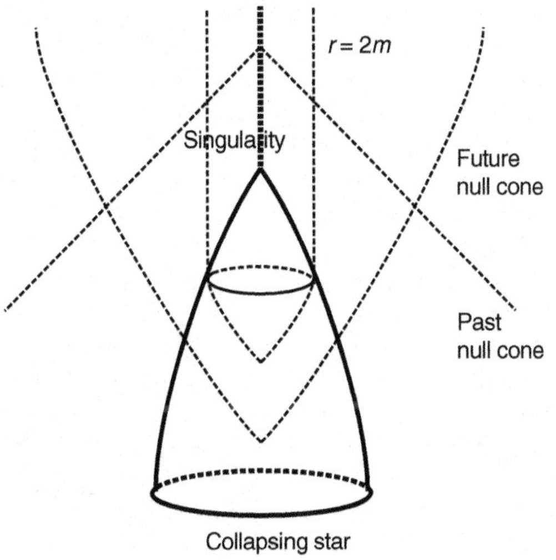

spherical body within the Schwarzschild radius must either be contracting of expanding. In our case we have contraction and hence $\dot{r}_0 < 0$. Since w^a is future-pointing and $\dot{r}_0 = w^a r_a < 0$, r^a must be past-pointing, and hence the star has entered region II of the extended Schwarzschild spacetime.

The absolute *minimum* rate of contraction is when $a = 0$, which gives

$$\dot{r}_0 = \sqrt{\frac{2m - r_0}{r_0}}.$$

By integrating this equation we see that

$$2\pi m = \int_{2m}^{0} \sqrt{\frac{r}{2m - r}}\, dr,$$

and hence the radius of any star will contract to zero, leaving a singularity, within proper time $2\pi m$ after it has crossed the Schwarzschild radius.

Figure 14.2 shows the exterior, matter-free region as part of a Kruskal diagram. The thick curve represents the surface of the star, and the region to its right represents the matter-free region. The region to its left is merely a mathematical construct representing the maximal extension of the matter-free part. In the actual situation this should be replaced by the spacetime of the interior. Note that once the star falls below the horizon, ($r = 2m$) and enters region II, an outgoing null ray cannot escape to infinity but falls into the singularity.

In a collapse situation of this type, the **event horizon** is defined to be the null surface that, in the matter-free region, is given by $r = 2m$. By

Figure 14.2. The thick curve represents the surface of the star. The region to its right represents the matter-free region.

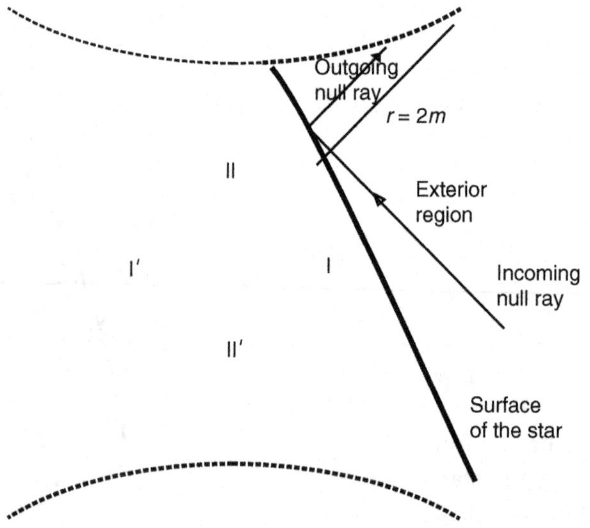

continuing this surface backwards in time, we encounter the source, after which it focuses to a point in the interior. The focusing behavior of the horizon in the interior is a consequence of the null Raychaudhuri equation with a source term $R_{ab}l^a l^b < 0$. The region B contained within the horizon is called a **black hole**. This is because it is completely invisible from the point of view of future null infinity: a future-directed curve through a point of B always stays within B and cannot escape to future null infinity. This is illustrated in Fig. 14.3.

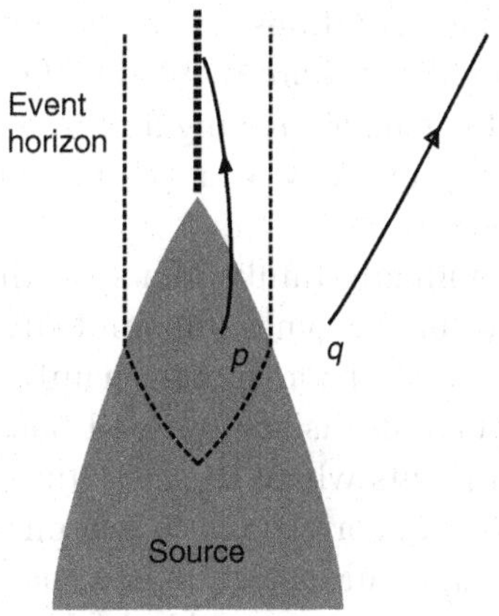

Figure 14.3. A future-directed curve starting from a point p within the horizon will always remain within the horizon and eventually fall into the singularity. A future-directed curve starting from a point q outside the horizon can escape to future null infinity.

As we showed earlier, a spherical body with radius less than $2m$ must either contract (i.e., it lies in region II) or expand (i.e., it lies in region II′). The choice between these two possibilities depends on how the body got to be where it is in the first place, that is, its previous history. In the case considered above, the body started contracting with a radius greater than $2m$, fell through the horizon, and then, inexorably, continued to contract to a singularity. There exists, however – at least in principle – the possibility of finding a body with radius less than $2m$ that is *expanding*. Though highly unlikely from a physical point of view, general relativity in itself does not exclude such a possibility. In this case, the body will, inexorably, continue to expand, at least until its radius is greater than $2m$. Moving backwards in time, it will contract until it forms a singularity. We thus have the bizarre situation of a singularity that, for no apparent reason, explodes to form a spherical body.

14.2 Singularities

Can the description of gravitational collapse leading to a singularity described in the previous section be regarded as realistic? An immediate objection is that it is based on an idealized spherically symmetric model and no object in nature is perfectly spherical. It may thus be the case that deviations from spherical symmetry may well prevent the formation of a singularity. However, perturbation calculations based on a Schwarzschild spacetime indicate that small deviations from spherical symmetry are rapidly radiated away and do not affect the overall description of spherical collapse. More importantly, the singularity theorems of Hawking and Penrose show that, under certain physically reasonable conditions, which do not require spherical symmetry, spacetime singularities are inevitable. One of the purposes of this chapter is to give the reader a feeling for these theorems (we do not include proofs) and show how they lead to a more realistic description of gravitational collapse.

One normally thinks of a spacetime singularity as a region in which the curvature becomes unboundedly large. However, as we have defined a spacetime to be a smooth manifold M together with a smooth metric g, the curvature is smooth and bounded at all points. We have simply omitted points where the curvature becomes unbounded, because such points do not correspond to real physical events.

As part of our definition of a spacetime we also demanded that (M, g) be maximal, or *inextendible,* in the sense that it cannot be embedded in a larger spacetime. This ensures that no physical event is omitted from M. We also define a smooth curve γ to be inextendible if there does not exist another smooth curve γ' such that $\gamma \cup \gamma'$ is also a smooth curve.

But, as we saw in the case of a Schwarzschild spacetime, inextendibility does not imply that an affine parameter along a null geodesic or proper time along a world line can assume an arbitrarily large value. For example, taking r to be the affine parameter, all radial null rays in a Schwarzschild spacetime have an affine parameter range of $(0, \infty)$, rather than $(-\infty, \infty)$ as in flat spacetime. The reason why r cannot assume a negative value is that there is simply no point of M such that $r = 0$. Such points can be added to M, considered as an abstract manifold, to make a larger space, $\hat{M}$ say, but the metric cannot be extended to $\hat{M}$.

Unfortunately, the addition of a singular points to a general spacetime manifold is a difficult and, perhaps, thankless undertaking. We therefore simply say that a spacetime is singular if it contains **holes** that could, in principle, be filled with singular points. Such holes make their presence felt by causing a geodesic to come to a full stop and thus limiting the range of its affine parameter. This leads to the definition of a singular spacetime:

Definition 14.1 A spacetime is said to be **singular** if there exists an inextendible null geodesic whose affine parameter range is not the whole real line, or an inextendible world line whose proper-time range is not the whole real line. Such curves are said to be **incomplete**.

While this definition does not fully capture the intuitive notion of singularity as a region where the curvature is unbounded, it does reflect the most objectionable feature of singularities, that there can be particles whose history can have a beginning or an end at some finite time. There are examples of spacetimes where this definition holds but where the curvature remains bounded, but it is thought that generically the curvature will diverge along geodesics whose affine parameter range is bounded in either the past or the future.

Let us now state our first singularity theorem, first proposed and proven by Penrose in 1965:

Theorem 14.1 *If the following conditions hold, a spacetime*[†] *is singular:*

(i) $R_{ab}l^a l^b \leq 0$ *for each future-pointing null vector* l^a.
(ii) *There exists a trapped surface.*
(iii) *There exists a noncompact, global Cauchy hypersurface, that is, the spacetime is predictable.*

The first condition expresses the fact that energy is locally positive and also that gravity is attractive. This condition is in fact contained in our definition of a physically reasonable spacetime. The second condition states that there exists a compact, two-dimensional surface having the property that both sets of future-pointing orthogonal null geodesics (outgoing and ingoing) converge in that both their divergences, ρ and ρ', are positive. As we have seen, this is true for the surface of a spherical collapsing star as soon as $r < 2m$. In fact, a theorem of Schoen and Yau (1983) asserts that a trapped surface must form whenever a sufficiently large quantity of matter is compacted into a sufficiently small region. This result has the advantage of not requiring spherical symmetry. The third condition is certainly the strongest of the three, but subsequent singularity theorems have been able to get away with a form of nonglobal predictability, which is more physically reasonable. The universe certainly seems predictable locally (at least classically), but we have no good reason to believe that this is true globally.

While this theorem was designed for predicting the occurrence of a singularity for gravitational collapse, our next theorem (proposed not long after by Hawking) has a more cosmological flavor.

Theorem 14.2 *If the following conditions hold, spacetime is singular in the past*

(i) $R_{ab}l^a l^b \leq 0$ *for each future-pointing timelike vector* l^a.
(ii) *There exists a spacelike global Cauchy hypersurface,* $t = t_0$ *say, on which the divergence function* θ *is strictly negative.*

As we shall see when we come to consider cosmological spacetimes, $\theta < 0$ on $t = t_0$ means that the universe is expanding at time $t = t_0$. The

[†] We define a spacetime to be a four-dimensional manifold M together with a metric g (of the appropriate signature) such that (M, g) is inextendible, has a space and a time orientation, and satisfies the dominant energy condition.

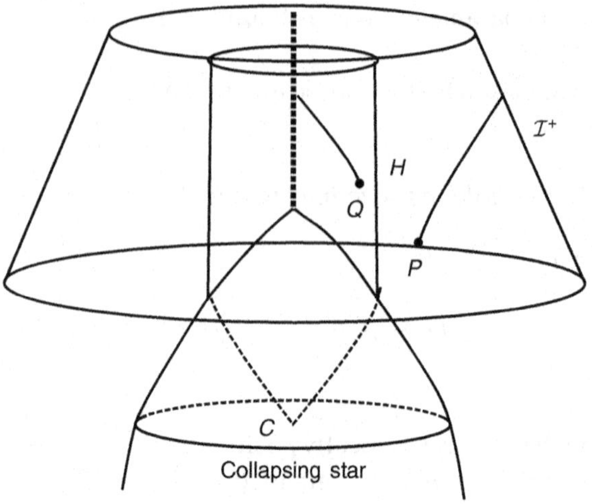

Figure 14.4. A world line from a point P outside the horizon can escape to infinity, $\mathcal{I}^+$, but a world line from a point Q within the horizon falls into the singularity.

theorem can therefore be interpreted as showing that if the universe is expanding in some epoch (e.g. now), then it must have been in some singular state (the big bang) at some finite time in the past. Again, the same conclusion can be reached under less stringent conditions, but for simplicity we shall not consider such refinements.

The startling thing about these two theorems and their subsequent refinements is that they show that singularities are *inevitable* for gravitational collapse once a trapped surface has formed, and also for an expanding universe.

14.3 Black Holes and Horizons

For spherical collapse we have seen that the singularity is always "clothed" by an event horizon H. World lines starting from points outside H can escape to future null infinity, $\mathcal{I}^+$, but those starting from points within H inevitably fall into the singularity. In a rescaled picture this is illustrated in Fig. 14.4.

Since singularities are also produced for nonspherical collapse (given that a trapped surface has been formed), it seems reasonable to suppose that the same will occur for a general situation, as long as we do not deviate too far from spherical symmetry.

Denoting by $I^-(\mathcal{I}^+)$ the set of points from which escape to $\mathcal{I}^+$ is possible, we define the event horizon as the boundary of $I^-(\mathcal{I}^+)$ [i.e. $H = \dot{I}^-(\mathcal{I}^+)$] and the black-hole region as $M - I^-(\mathcal{I}^+)$. The reason for these definitions is, of course, that they give what we want in the spherical case. Furthermore, some crucial properties of H follow directly from defining it in this way. We summarize these in the following theorem [see, for example, Wald (1984)]:

Theorem 14.3 *Each point of H lies on a future-inextendible null geodesic segment that lies entirely in H, and the divergence ρ of these null generators of H cannot become infinite at any point of H.*

A null geodesic may thus enter H at some point (for example, the caustic point C in Fig. 14.4), but thereafter it must remain in H and never leave it. This leaves open the possibility of several entry points, corresponding to a situation involving several black holes. A simple consequence of

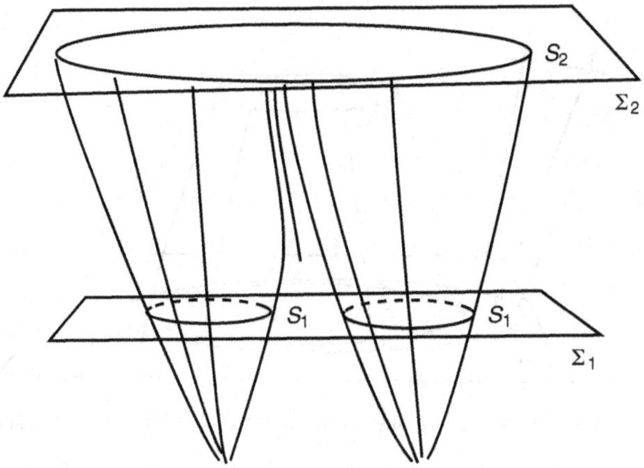

Figure 14.5. The fusion of two black holes.

the theorem is that they may fuse together but never thereafter separate (Fig. 14.5).

We are now in a position to state one of the most important theorems concerning black holes:

Theorem 14.4 *The area of a cross section of H increases in future directions. More precisely, if Σ_1 and Σ_2 are two spacelike hypersurfaces, Σ_2 to the future of Σ_1, that intersect H in two cross sections, S_1 and S_2, then the area of S_2 is greater than or equal to that of S_1.*

Unfortunately, we do not have the necessary mathematical machinery to give a full, rigorous proof of this theorem. Instead, we shall illustrate the main ideas in the simple case of a single black hole.

We first show that $\rho \leq 0$, where ρ is the divergence of the null generators of H. Suppose this were not the case and there existed a point p on a cross section S such that $\rho(p) > 0$. We could then deform S outward in a neighborhood of p to get a two-surface S' containing a point $p' \in I^-(\mathcal{I}^+)$ and a pencil of null rays orthogonal to S' that escaped to infinity and whose divergence satisfied $\rho(p') > 0$ (see Fig. 14.6). The null Raychaudhuri equation,

$$D\rho = \rho^2 + \sigma\bar{\sigma} - \tfrac{1}{2} R_{ab} l^a l^b,$$

would then imply that $\rho > 0$ on the null ray through p'. This, however, gives a contradiction, since ρ must tend to zero as we approach infinity.

Since $\rho \leq 0$ on H, equation (11.4), namely $DA = -2\rho A$, shows that an area element A of S increases in future directions and hence that the total area of S increases as we move up H.

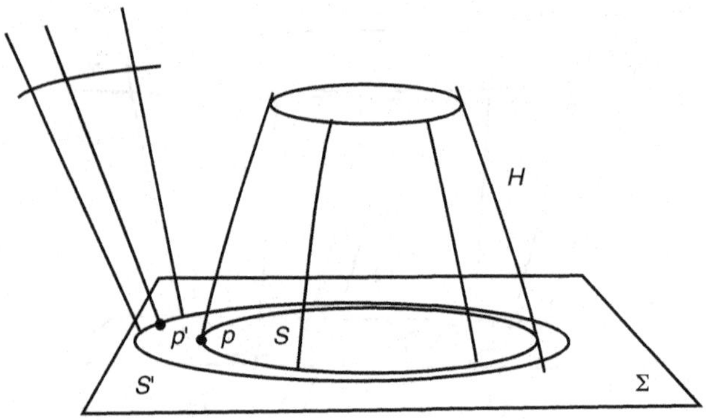

Figure 14.6. Null geodesics orthogonal to the outwardly distorted cross section, S' can escape to infinity.

So far we have tacitly assumed that all singularities that arise in gravitational collapse lie within an event horizon and thus cannot be seen from null infinity, exactly as in spherical collapse. This is, perhaps, a safe assumption to make for small deviations from spherical symmetry, but is it true in general? Though no rigorous proof exists at the moment, there is a considerable body of indirect evidence that indicates that this is indeed the case. This has led Penrose to make his famous **cosmic censorship hypothesis,** which states that gravitational collapse cannot lead to naked singularities, or, in other words, singularities that are not clothed by an event horizon. Counterexamples to cosmic censorship do exist, but they all make assumptions that, in one way or another, are physically unacceptable. The consensus now seems to be that collapse involving physically reasonable matter cannot produce naked singularities.

14.4 Stationary Black Holes and Kerr Spacetime

Consider a body that undergoes complete gravitational collapse and forms a black hole. In the spherical case, the spacetime outside the black hole will have a Schwarzschild geometry. However, in the nonspherical case, the spacetime geometry outside the collapsing body should vary with time and depend on the details of the collapse. Nevertheless, one would expect on physical grounds that at sufficiently late times the spacetime geometry should settle down to a stationary state and that all matter should eventually fall into the black hole, leaving a stationary vacuum region outside the horizon. It is therefore of great interest to determine the geometry of this region.

Though there are certain details still to be resolved, a string of results, which have become known as the **no-hair theorem**, show that the vacuum region outside a stationary black hole has a Kerr geometry, and is thus determined by just two numbers, the mass m and the angular momentum parameter a. All other degrees of freedom are simply swallowed up by the black hole, leaving a particularly clean exterior geometry depending on only two parameters.

Since the proof of the no-hair theorem is very complicated, we shall not attempt to reproduce it here. However, as its consequences are so profound, it is important to gain some intuitive feeling for why it should be true. It can be shown without too much difficulty that a stationary perturbation of a Kerr black hole leads to nothing more exciting than another Kerr black hole. Thus, by a sequence of such perturbations we can always arrive at a Schwarzschild black hole (Kerr with $a = 0$), and, conversely, starting with a Schwarzschild black hole, we can never leave the vicious circle of Kerr black holes. Starting with a spherical collapse situation leading to a Schwarzschild black hole, we can make a sequence of stationary perturbations by, for example, throwing in bits of matter and waiting for the black hole to settle down to a stationary state. But since we cannot leave the vicious circle of Kerr black holes, this will always lead to a Kerr black hole. This argument is not, of course, watertight, but it does give one a feeling as to why the no-hair theorem is true. It also has a thermodynamic flavor to it, a point which we shall return to shortly.

In Section 12.3 we defined a Kerr spacetime to be stationary, axisymmetric, asymptotically flat, and algebraically special, and showed the following:

(i) There exists a preferred radial coordinate r and a preferred latitudinal angle θ such that

$$\theta_a\theta^a = -\Sigma^{-1} \quad \text{and} \quad r_a\theta^a = 0, \tag{14.1}$$

where $r_a = \nabla_a r$, $\theta_a = \nabla_a\theta$, and $\Sigma = r^2 + a^2\cos^2\theta$.

(ii) There exist two shear-free, affine, null vector fields, l_a (future-pointing) and l'_a (past-pointing), such that

$$l_a r^a = 1, \qquad l_a\theta^a = 0, \qquad l_a\xi^a = 1, \qquad l_a\chi^a = -a\sin^2\theta, \tag{14.2}$$

$$l'_a r^a = 1, \qquad l'_a\theta^a = 0, \qquad l'_a\xi^a = -1, \qquad l'_a\chi^a = a\sin^2\theta. \tag{14.3}$$

As in the Schwarzschild case, l_a and l'_a are called **principal null vectors**. All geometric quantities are, of course, preserved under the symmetries in that they are killed by $\mathcal{L}_\xi$ and $\mathcal{L}_\chi$.

In Section 12.3 we also showed that two ignorable coordinate functions, u and φ, can be introduced such that, in terms of coordinate system (u, r, θ, φ),

$$l^a = \partial^a_r, \qquad \xi^a = \partial^a_u, \quad \text{and} \quad \chi^a = \partial^a_\varphi \tag{14.4}$$

and

$$\begin{aligned} g = {} & \left(1 - \frac{2mr}{\Sigma}\right) du^2 + 2\, du\, dr + \frac{2mr}{\Sigma} 2a \sin^2\theta\, du\, d\varphi \\ & - 2a\sin^2\theta\, dr\, d\varphi - \Sigma\, d\theta^2 \\ & - \left((r^2 + a^2)\sin^2\theta + \frac{2mr}{\Sigma} a^2 \sin^4\theta\right) d\varphi^2. \end{aligned} \tag{14.5}$$

The complexity of this form of the metric is partially due to the introduction of nongeometric structure in the form of the coordinate functions u and φ. Fortunately, things will appear less complex when we extract its geometric content.

Using equations (14.5) and (14.4), we obtain

$$\xi^2 = \xi_a \xi^a = g(\partial_u, \partial_u) = 1 - \frac{2mr}{\Sigma}, \tag{14.6}$$

$$\xi\chi = \xi_a \chi^a = g(\partial_u, \partial_\varphi) = \frac{2mr}{\Sigma} a \sin^2\theta, \tag{14.7}$$

$$\chi^2 = \chi_a \chi^a = g(\partial_\varphi, \partial_\varphi) = -(r^2 + a^2)\sin^2\theta - \frac{2mr}{\Sigma} a^2 \sin^4\theta. \tag{14.8}$$

Similarly,

$$\xi_a = g_{ab}\xi^b = \xi^2 u_a + r_a + \xi\chi\varphi_a, \tag{14.9}$$

$$\chi_a = g_{ab}\chi^b = \xi\chi u_a - 2a\sin^2\theta\, r_a - \chi^2\varphi_a, \tag{14.10}$$

$$l_a = g_{ab} l^b = u_a - a\sin^2\theta\, \varphi_a. \tag{14.11}$$

From these equations we obtain the geometric relation

$$l_a = \frac{1}{\Delta}[(r^2 + a^2)\xi_a + a\chi_a - \Sigma r_a], \tag{14.12}$$

where $\Delta = r^2 - 2mr + a^2$. Furthermore, by TP invariance, we also have

$$-l'_a = \frac{1}{\Delta}[(r^2 + a^2)\xi_a + a\chi_a + \Sigma r_a]. \tag{14.13}$$

Two more important geometric equations can be obtained by contracting equation (14.12) with l'^a and with r^a. This gives

$$l'_a l^a = -2\frac{\Sigma}{\Delta}, \tag{14.14}$$

$$r_a r^a = -\frac{\Delta}{\Sigma}. \tag{14.15}$$

A slightly longer calculation gives

$$C_{abcd}C^{abcd} = 48\frac{m^2}{\Sigma^3}, \tag{14.16}$$

which shows that a Kerr spacetime is singular on the equatorial, ring-like region $r=0$, $\theta=\pi/2$. Note that all the above equations reduce to the corresponding Schwarzschild versions for $a=0$.

Energy and Angular Momentum

Let us now use Definition 12.3 to obtain expressions for the total energy and intrinsic angular momentum of a Kerr spacetime in terms of the parameters a and m.

For the two-surface S, we use the intersection of the two three-surfaces u = const and r = const. The next step is to find an orthogonal pair of unit vectors that are orthogonal to S. Since

$$\chi^a u_a = \theta^a u_a = \chi^a r_a = \theta^a r_a = 0,$$

the vectors χ^a and θ^a are *tangent* to S. Thus, since $r^a\chi_a = r^a\theta_a = 0$, the vector r^a is *orthogonal* to S. Furthermore, the vector

$$v^a = \xi^a - \frac{\xi\chi}{\chi^2}\chi^a$$

satisfies

$$v^a\chi_a = v^a\theta_a = v^a r_a = 0$$

and is therefore orthogonal to S as well as being orthogonal to r^a. Since r^a and v^a are unit vectors asymptotically, Definition 12.3 gives

$$E = -\lim_{r\to\infty}\frac{1}{4\pi}\int_S \nabla_a\xi_b v^{[a}r^{b]}\,dS, \tag{14.17}$$

$$J = \lim_{r\to\infty}\frac{1}{8\pi}\int_S \nabla_a\chi_b v^{[a}r^{b]}\,dS, \tag{14.18}$$

where we note that $r^{-2}\,dS \to \sin\theta\,d\theta\,d\varphi$.

Using equation (14.9), we have

$$\begin{aligned}(v^a r^b - r^a v^b)\nabla_a\xi_b &= (v^a r^b - r^a v^b)(u_b\nabla_a\xi^2 + \varphi_b\nabla_a\xi\chi)\\ &= -r^a\nabla_a\xi^2 + \frac{\xi\chi}{\chi^2}r^a\nabla_a\xi\chi\\ &= 2mr^{-2} + O(r^{-3}),\end{aligned}$$

and this immediately gives

$$E = m. \tag{14.19}$$

Similarly, using equation (14.10), we have

$$\begin{aligned}(v^a r^b - r^a v^b)\nabla_a \chi_b &= (v^a r^b - r^a v^b)(u_b \nabla_a \xi\chi - a r_b \nabla_a \sin^2\theta + \varphi_b \nabla_a \chi^2) \\ &= -r^a \nabla_a \xi\chi + \frac{\xi\chi}{\chi^2} r^a \nabla_a \chi^2 \\ &= -6mar^{-2}\sin^2\theta + O(r^{-3}),\end{aligned}$$

which, when substituted in equation (14.18), gives

$$J = am. \tag{14.20}$$

The Killing Tensor and Geodesics

A symmetric tensor field K_{ab} satisfying

$$\nabla_{(a}K_{ab)} = 0$$

is called a **Killing tensor**. A remarkable property of a Kerr spacetime is that it admits a Killing tensor given by

$$K_{ab} = \Delta l_{(a} l'_{b)} + r^2 g_{ab} \tag{14.21}$$

(Walker and Penrose 1970). That K_{ab} given by this equation satisfies $\nabla_{(a}K_{ab)} = 0$ can easily be checked using equations (14.13) and (14.14).

The existence of a Killing tensor greatly facilitates the calculation of orbits in a Kerr spacetime. Let u^a be an affine tangent vector to a geodesic γ. Then, by Theorem 12.1, $E = \xi_a u^a$ and $L = -\chi_a u^a$ are constant on γ. Moreover, by an obvious extension of this theorem, $K = K_{ab}u^a u^b$ is also constant on γ. We thus have *four* constants of motion: $u_a u^a$, which is either 0 or 1 depending on whether γ is null or timelike, and E, L, and K. Using the same method as in the Schwarzschild case, the equation of motion of γ can now be written down in terms of these constants of motion.

The Horizon

As in the Schwarzschild case, we define the event horizon by $r_a r^a = 0$ (we justify this shortly). Since we are primarily interested in gravitational collapse, where, according to cosmic censorship, the singularity must be surrounded by an event horizon, the equation

$$r_a r^a = -\frac{1}{\Sigma}(r^2 - 2mr + a^2) = 0$$

must have positive real roots, and hence $m \geq |a|$.

By starting at past null infinity and travelling into the interior along a principal null direction given by l'_a, we first encounter the event horizon H at

$$r = r_H = m + \sqrt{m^2 - a^2}.$$

Since l'_a is past-pointing and $l'_a r^a = 1$, the vector r^a, which is null on H, is past-pointing on H. That H is in fact the event horizon in the sense of a one-way membrane can now be seen as follows. Let v^a be the four-velocity of a particle crossing H. Since v^a is future-pointing and r^a is past-pointing, $v^a r_a = v^a \nabla_a r < 0$, and hence r must necessarily *decrease* as the particle crosses H.

We note that, as in the Schwarzschild case, l_a blows up near H. This is simply a reflection of the fact that future null infinity, $\mathcal{I}^+$, is inaccessible from H. If l_a were well defined on H, then we could gain access to $\mathcal{I}^+$ by travelling up its rays. Like a Schwarzschild spacetime, a global Kerr spacetime possesses two types of event horizon: a black-hole horizon where l'_a is well defined and r^a is past-pointing, and a white-hole horizon where l_a is well-defined and r^a is future-pointing. Here we are only interested in the former.

Since l'_a is well defined on H where $\Delta = 0$, equation (14.13) implies

$$(r_H^2 + a^2)\xi_a + a\chi_a + r_a = 0 \tag{14.22}$$

on H. Thus

$$\Psi^a = \xi^a + \Omega\chi^a, \tag{14.23}$$

where

$$\Omega = \frac{a}{r_H^2 + a^2}, \tag{14.24}$$

is a Killing vector field that is tangent to the null generators of H. As we shall now show, the number Ω gives the **angular velocity** of the black hole.

We must first, however, define what we mean by the angular velocity of a Kerr black hole. Let φ be a longitudinal angle satisfying $\chi^a \nabla_a \varphi = 1$ and subject to the following two conditions:

(i) $\Psi^a \nabla_a \varphi = 0$, that is, φ is constant along the null generators of H.
(ii) $l'^a \nabla_a \varphi = 0$, that is, φ is constant along the rays of l'^a and hence well defined near $\mathcal{I}^-$.

Given such a φ, we define the angular velocity to be $\Omega = -\dot{\varphi} = -\xi^a \nabla_a \varphi$ near $\mathcal{I}^-$. If, for example, the black hole has angular velocity Ω and a stationary observer near $\mathcal{I}^-$ directs two flashes of light with time interval Δt along a principal null direction, then they will hit the surface H

of the black hole with φ-interval given by $\Delta\varphi = \Omega\Delta t$. That Ω given by equation (14.24) satisfies this definition can be seen as follows: Since $\Psi^a\nabla_a\varphi = 0$ and $\chi^a\nabla_a\varphi = 1$, equation (14.23) gives $\Omega = -\xi^a\nabla_a\varphi$ on H. Furthermore, $l'^a\nabla_a\varphi = 0$ and $\mathcal{L}_\xi l'_a = 0$ can be used to show that $\xi^a\nabla_a\varphi$ is constant along the principal null directions and hence that $\Omega = -\xi^a\nabla_a\varphi$ is also valid near $\mathcal{I}^-$.

The remarkable thing about this result is that Ω is independent of θ, and hence the black hole rotates as a rigid body.

Even though r^a and Ψ^a are tangent to the null geodesic generators of H, they are *not* affine tangent vectors. This can be seen as follows: Using equation (14.15), we have

$$r^b\nabla_b r_a = r^b\nabla_a r_b = \frac{\nabla_a(r^b r_b)}{2} = -\frac{r_H + m}{\Sigma} r_a$$

on H. Also, equation (14.22) gives

$$r^a = -\frac{\Sigma}{r_H^2 + a^2}\Psi^a$$

and hence

$$\Psi^b\nabla_b\Psi_a = \kappa\Psi_a, \tag{14.25}$$

where

$$\kappa = \frac{m r_H}{r_H^2 + a^2}. \tag{14.26}$$

The remarkable thing about this result is that κ, which is called the **surface gravity** of the black hole and which is a measure of the nonaffineness of Ψ_a, is, like Ω, independent of θ.

Since H is a null surface, it has a well-defined divergence,

$$\rho = m^a\bar{m}^b\nabla_a\Psi_b,$$

which is real, and a well-defined shear, $\sigma = m^a m^b\nabla_a\Psi_b$. That these quantities are in fact zero follows trivially from the Killing equation $\nabla_a\Psi_b + \nabla_b\Psi_a = 0$.

Since $\rho = 0$, an argument similar to that used in Section 12.1 shows that all spacelike cross sections of H have the same area A. In order to find an expression for A we use χ^a and θ^a. Since

$$\chi^a r_a = \theta^a r_a = \mathcal{L}_\chi\theta = \chi^a\theta_a = 0,$$

they are mutually orthogonal and tangent to H. From equation (14.1) we have $\hat{\theta}_a\hat{\theta}^a = -\Sigma$ and $\hat{\theta}^a\nabla_a\theta = -1$, where $\hat{\theta}^a = \Sigma\theta^a$. Furthermore, equation (14.8) implies

$$\chi_a\chi^a = -(2mr_H\sin\theta)^2/\Sigma$$

on H. We thus see that $\hat{\theta}^a \nabla_a \theta = -1$, where $\hat{\theta}^a$ has length $\sqrt{\Sigma}$, and $\chi^a \nabla_a \varphi = 1$, where χ^a has length $(2mr_H \sin\theta)/\sqrt{\Sigma}$. This implies that the area spanned by $\hat{\theta}^a\,\delta\theta$ and $\chi^a\,\delta\varphi$ is $2mr_H \sin\theta\,\delta\theta\,\delta\varphi$, and hence that the total area is given by

$$A = 2mr_H \iint \sin\theta\, d\theta\, d\varphi = 8\pi m r_H. \tag{14.27}$$

We have now shown that a Kerr black hole has energy $E = m$ and intrinsic angular momentum $J = ma$, as well as having angular velocity Ω, surface gravity κ, and area A given by equations (14.24), (14.26), and (14.27) – all of which are constants and which depend only on m and a.

By making small increments, δa and δm, in a and m, we arrive at the remarkable equation

$$\delta E = \frac{1}{8\pi}\kappa\,\delta A + \Omega\,\delta J, \tag{14.28}$$

which is one of the key results in black-hole thermodynamics, a topic we shall consider shortly. For now we merely note that if $\delta A = 0$, then this equation reduces to $\delta E = \Omega\,\delta J$, which is identical to the corresponding Newtonian equation for a rigid rotating body. For example, if we take a single particle with mass m, and angular velocity Ω whose distance from the axis of rotation is some fixed number r, then $E = mr^2\Omega^2/2$ and $J = mr^2\Omega$, and hence $\delta E = mr^2\Omega\,\delta\Omega = \Omega\,\delta J$.

14.5 The Ergosphere and Energy Extraction

Equation (14.6) may be expressed as

$$\xi_a \xi^a = \frac{1}{\Sigma}(r^2 - 2mr + a^2\cos^2\theta). \tag{14.29}$$

Thus $\xi_a\xi^a < 0$, and hence ξ^a is spacelike, in the region where

$$r_H \le r < r + \sqrt{m^2 - a^2\cos^2\theta}.$$

This region, which is called the **ergosphere**, lies outside the horizon except at the poles where the horizon and the surface of the ergosphere touch. Since ξ^a is spacelike, no stationary observer can exist in the ergosphere even though it lies outside the black hole.

Since a black hole is a region from which no matter can escape, not even light, it came as a great surprise when Penrose (1969) proved that energy can be extracted from a black hole with an ergosphere. The proof is based on the following two key facts:

(i) If $p^a = m_0 v^a$ is the four-momentum of a geodesic particle then, by Theorem 12.1, $E = \xi^a p_a$ is constant along its world line. Furthermore, if the particle escapes to infinity, where ξ^a tends to the four-velocity of a stationary observer, then E is equal to the particle's energy with respect to this observer. Similarly, $L = -\chi^a p_a$ is constant along the world line, and if the particle escapes to infinity, L is equal to the particle's orbital angular momentum with respect to a stationary observer at infinity. (The minus sign in the definition of L is because of the signature $(+, -, -, -)$ of our metric.)

(ii) Since ξ^a is spacelike in the ergosphere, there can exist perfectly reasonable particles with positive rest mast, m_0, and future-directed four-velocity v^a such that $E < 0$. Such particles cannot, of course, leave the ergosphere (outside the ergosphere, ξ^a is future-directed and hence $E > 0$), but either remain in it forever in a closed orbit or fall into the black hole.

With these two facts in mind, consider a particle with energy $E_0 = \xi_a p_0^a$ that falls from infinity into the black hole (see Fig. 14.7).

Suppose that when inside the ergosphere it breaks into two fragments, one of which has *negative* energy $E_1 = \xi_a p_1^a$. By conservation of four-momentum (i.e., $p_0^a = p_1^a + p_2^a$), the energy $E_2 = \xi_a p_2^a$ of the other fragment will be given by $E_2 = E_0 - E_1$, and will be *greater* than the initial energy E_0. One can explicitly verify that the breakup can be arranged so that the fragment with negative energy falls into the black hole while the other fragment with energy $E_2 > E_0$ escapes to infinity. Thus, by means of this process, an observer at infinity can extract energy $E_2 - E_0$ from the black hole. The black hole has essentially absorbed the negative energy E_1, causing $E_2 - E_0$ to be positive.

At first sight it might appear that by absorbing a negative amount of energy the area of the black hole will decrease, in contradiction to the area theorem. However, as we shall now show, the black hole also absorbs a

Figure 14.7. Energy extraction from a Kerr black hole. The outgoing energy E_2 is greater than the incoming energy E_0. A particle with negative energy E_1 is absorbed by the black hole.

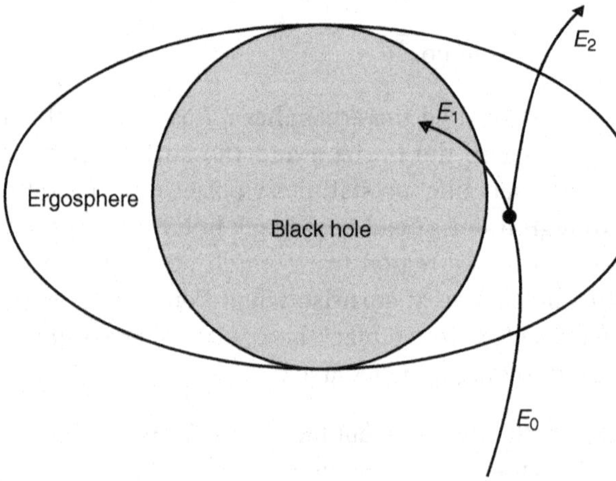

negative amount of angular momentum (angular momentum of the opposite sign to that contained in the black hole) and this conspires to cause the area to increase. This not only is in accord with the area theorem, but sets a limit to the amount of energy that can be extracted. By absorbing negative angular momentum the black hole will eventually become Schwarzschild from which no further energy can be extracted because of the lack of an ergosphere.

Suppose, in the above process, the black hole absorbs energy $\delta E = \xi_a p^a$ and angular momentum $\delta J = -\chi_a p^a$, where δE is negative. Then, since $\Psi^a = \xi^a + \Omega\chi^a$ is future-pointing on H, we have

$$\delta E - \Omega\,\delta J > 0,$$

and hence, taking Ω to be positive, δJ must be negative. Furthermore, by equation (14.28) we have

$$\frac{1}{8\pi}\kappa\,\delta A = \delta E - \Omega\,\delta J > 0,$$

which shows that the area increases.

14.6 Black-Hole Thermodynamics

The reader will have noticed that much of what we have said about the formation and final state of black holes has a strong thermodynamical flavor. For example, a star that undergoes complete gravitational collapse has, initially, many degrees of freedom, but when it eventually settles down to steady Kerr state its degrees of freedom are reduced to just two parameters, m and a. Similarly, a thermal system – for example, a liquid globule in an orbiting space station – may have an initial state containing infinitely many degrees of freedom, but, due to internal friction, these are rapidly reduced to just two (macroscopic) degrees of freedom – for example, total kinetic energy and total angular momentum – when it attains thermal equilibrium. Furthermore, the area of a black hole increases with time until it reaches a Kerr state, just as the entropy of an isolated thermal system increases until it achieves thermal equilibrium.

The analogy between black-hole physics and thermodynamics is further strengthened by equation (14.28), namely

$$\delta E = \frac{1}{8\pi}\kappa\,\delta A + \Omega\,\delta J, \tag{14.30}$$

which bears a striking resemblance to the key thermodynamical relation

$$\delta E = T\,\delta S - P\,\delta V. \tag{14.31}$$

Indeed, a simple rotating thermal system, such as a liquid globule in thermal equilibrium, satisfies the equation

$$\delta E = T\,\delta S + \Omega\,\delta J$$

where Ω is its angular velocity and J its intrinsic angular momentum in an equilibrium state.

But should we take the similarity between black-hole physics and thermodynamics as more than a mere analogy? After all, black holes and thermal systems are very different creatures. The fact that the area of a black hole increases with time follows rigorously from general relativity and cosmic censorship, whereas the increase of entropy for a thermal system is, ultimately, a statistical effect. Also, the degrees of freedom contained in the initial state of a collapsing star are irretrievably lost in the black hole, whereas the microscopic degrees of freedom contained in a thermal system in equilibrium still exist but have little effect on its statistical (macroscopic) degrees of freedom, such as temperature and pressure.

The first real indication that the relationship between black-hole physics and thermodynamics is more than a mere analogy came with the discovery by Hawking (1975) that a black hole actually radiates like a blackbody of temperature $\kappa/2\pi$. This is called the **Hawking effect** and is a consequence of the behavior of *quantum* fields in the presence of an event horizon. It should be emphasized that this is a quantum effect – essentially the first fruit of putting quantum theory together with general relativity – and does not contradict the fact that a black hole is indeed black, classically.

According to the Hawking effect, the quantity $\kappa/2\pi$, which was previously thought to be only analogous to temperature, actually *is* the temperature of a black hole. On comparing equations (14.30) and (14.31), this suggests that we should take $A/4$ to be the black hole's entropy. But, apart from the fact that the area of a black hole is a geometric rather than a statistical quantity, this immediately presents us with a difficulty. By radiating according to the Hawking effect the black hole will lose energy, resulting in a *decrease* in its area and hence in its entropy. But the thermal radiation produced by this process will tend to increase the entropy of matter outside the black hole. Again, being a quantum effect, this does not contradict the (classical) black-hole area theorem. It does, however, relate to a corresponding difficulty with classical thermodynamics in the presence of a black hole. On throwing matter into a black hole, its entropy vanishes into the singularity, resulting in a net decrease in the entropy S_M of the total matter of the universe, but this will also result in an increase of the black hole's area. Thus, although S_M and $A/4$ can vary individually, it is possible that the **generalized entropy** S defined by

$$S = S_M + \tfrac{1}{4}A \qquad (14.32)$$

never decreases. The conjecture that $\delta S \geq 0$ in all processes was first proposed by Bekenstein (1974) – prior to the discovery of the Hawking effect – and is known as the **generalized second law**. As with the cosmic censorship hypothesis, which is closely related to it, many attempts have been made to find a counterexample to the generalized second law, but no one as yet has succeeded. All indications are that it is, indeed, a true and very fundamental law of nature with very profound consequences for the structure of the universe as a whole.

EXERCISES

14.1 Starting from rest at infinity, an observer falls into a spherical black hole. As he crosses the horizon, he receives a radial photon that was transmitted from infinity with frequency ω_0. What is the received frequency?

14.2 A particle falls into a spherical black hole. From the point of view of a stationary observer at infinity, how long will it take to travel between $r = 3m$ and $r = 2m$?

14.3 For a Kerr spacetime,

$$\rho = \bar{m}^a m^b \nabla_a l_b = -\frac{1}{r + ia\cos\theta}.$$

Use this to show that

$$\nabla^a l_a = -\frac{2r}{\Sigma}.$$

14.4 Use the previous exercise to show that $\Box r^{-1} = 0$, where $\Box = \nabla_a \nabla^a$.

14.5 Assume that the exterior region of a rotating star has a Kerr geometry and $r =$ const on its surface. If the star emits light of a constant frequency, how will the star appear to an observer at infinity?

14.6 For a Kerr spacetime show that the angle ϕ between two principal null directions according to a stationary observer is given by

$$\cos\phi = 1 - 2\frac{r^2 - 2mr + a^2\cos^2\theta}{r^2 - 2mr + a^2}.$$

Note that $\phi = \pi$ for a Schwarzschild spacetime where $a = 0$.

14.7 Due to the shearing influence of the Weyl curvature, a small spherical object will appear as a distorted disk in the night sky. How will it appear when viewed along a principal null direction?

14.8 For a Schwarzschild spacetime ($a = 0$) show that K_{ab} given by equation (14.21) is a Killing tensor. Give a geometrical interpretation of K_{ab}.

14.9 For a timelike orbit with four-velocity w^a in a Schwarzschild spacetime, the quantities $E = w^a \xi_a$, $J = w^a \chi_a$ and $K = K_{ab} w^a w^b$ are constants of the motion. Show that $K = -J^2$ and hence does not give a new, independent constant of the motion.

PART FOUR

COSMOLOGY

15

The Spacetime of the Universe

We now turn to the spacetime description of the universe as a whole. At first sight this may seem like a formidable undertaking, but as we shall be interested in only gross, very large-scale features – a "point event" will contain many galaxies and extend for millions of years – it turns out to be quite tractable.

Due to the high degree of symmetry possessed by the universe on a suitably large scale, much of the mathematical machinery developed in the previous chapters (metric tensors, curvature tensors, etc.) is not strictly necessary to obtain an overall picture. Indeed, in this chapter, we shall not even use the spacetime metric, and simply content ourselves with the properties of photons and null rays. This, as we shall see, gives an adequate description of the causal properties of the universe, particularly as regards horizons. The full, relativistic treatment will be left to the next chapter.

15.1 The Cosmological Principle

Roughly speaking, the **cosmological principle** states that, at any given time, the universe looks the same to all observers in all typical galaxies (galaxies that do not have any large peculiar motion of their own, but are simply carried along with the general cosmic flow of galaxies), and in whatever direction they look. Clearly this principle is not true on a human scale – if it were true, the universe would be a pretty boring place. Even on a very large astronomical scale it is false. For example, our galaxy (the Milky Way) belongs to a small local group of other galaxies, which in turn lies near the enormous cluster of galaxies in Virgo. In fact, of the 33 galaxies in Messier's catalogue, almost half are in one part of the sky, the constellation Virgo. Thus, even on a scale where individual galaxies (or even clusters of galaxies) may be taken as point particles, the cosmological principle is false. The cosmological principle comes into play only when we view the universe on a scale at least as large as the distance between clusters of galaxies, or about 10^8 light years. In fact, most observational evidence (but not all!) suggests that the cosmological principle is true for a sufficiently large length scale but one that is small in comparison with the characteristic size of the universe. (This may be taken to be ct, where

t is the age of the universe.) In order to imagine what this means, it may be helpful to imagine a block of polystyrene of about one yard across. The block would appear inhomogeneous on a scale of about one-hundredth of an inch, but homogeneous on a scale of about one inch.

Should we be surprised that, at least on a sufficiently large scale, the cosmological principle is (apparently) true for the universe? Perhaps we should, especially when we consider how the universe *could* have been (consistent with the laws of physics). It could, for example, have been completely chaotic, with no apparent structure at any length scale. It could have been fractallike with a different (or the same) structures occurring at different length scales: clusters of galaxies, clusters of clusters, and so on. The creator (or whatever) obviously took great care to make the universe homogeneous.

The cosmological principle has, of course, great philosophical appeal. It states that at any given time there are no preferred points or directions in the universe. After all, why should one point or one direction be preferred to any other? One must, however, be wary of philosophical principles in physics. What counts is not how we would like the universe to be, but how it actually is, and this should be determined by observation. There is a stronger form of the cosmological principle, known as the **perfect cosmological principle,** which states that the universe should also look the same at all times. This has even greater philosophical appeal, but is almost certainly not true.

Even if we do not attach too much weight to philosophical principles, perhaps we should not be all that surprised that the universe is homogeneous. There are, after all, many physical systems that become homogeneous of their own accord – a gas in thermal equilibrium, for example. The approach to thermal equilibrium is, however, caused by interparticle interactions (e.g., collisions), and without such interactions thermal equilibrium would not in general occur. Without any interaction – for example, thermal contact – there would be no reason why two cylinders of gas should have the same temperature. A similar situation applies to the universe: there exist regions of the universe that have, according to the cosmological principle, the same density, temperature and so on, but which have never (even in principle) interacted with each other. It is rather like finding two cultures that have the same language but, as far as we know, have never been in contact with each other. That would certainly surprise us and demand some sort of explanation, as should the truth of the cosmological principle.

As we have seen, there are relatively small fluctuations from homogeneity in the present epoch. However, as we shall see, these fluctuations become smaller as we go back in time, and at very early times the universe was almost perfectly homogeneous – at least according to the standard theory. The problem thus becomes acute for the very early universe. There have been various attempts to address this problem (most notably inflationary theories) using known physical laws, but these have not been entirely successful and, it can be argued, are rather contrived. Another viewpoint is that the extreme homogeneity of the early universe is a reflection of some as yet unknown, very fundamental law of nature connected with quantum gravity, gravitational entropy, and the arrow of time.

Let us now give a more precise formulation of the cosmological principle in terms of a spacetime picture.

First isotropy. We demand that there exist a preferred set of observers, which we call **comoving**, to whom the universe appears isotropic and whose world lines form a congruence. By a **congruence** we mean a family of nonintersecting curves such that every point lies on one and only one curve. To such an observer the background radiation would, for example, appear isotropic. A terrestrial observer is not comoving in this sense. This is because of the motion of the earth around the sun and, more importantly, the peculiar motion of our galaxy toward the giant cluster of galaxies in constellation Virgo. Roughly speaking, a comoving observer is one who moves with the general cosmic flow of galaxies with no peculiar motion of his own.

Next homogeneity. We demand that to any point p on some preferred world line l, there exists a corresponding point p' on any preferred other world line l' such that the conditions at p are the same as those at p', according to comoving observers occupying these points. The points of the spacetime M thus fall into equivalence classes, which we call **epochs**. We assume that each epoch forms a smooth three-dimensional hypersurface in M. Two comoving observers will belong to the same epoch if their conditions are identical. Each one would, for example, find that the background radiation has the same temperature. If all comoving observers set their clocks to zero in some epoch, the present one for instance, we obtain a function t on M where $t(p)$ is the reading on the observer's clock who instantaneously occupies p. This function is called **universal time**. Note that t is a constant on each epoch. If this were not so, then there would exist comoving observers in the same epoch with different readings on their clocks. Their conditions would thus not be identical, and they would not belong to the same epoch.

15.2 Cosmological Red Shifts

Consider two nearby, comoving particles that at the present time, $t = t_0$, are at some standard distance apart, say one light-second. The time it takes light to make the return journey, there and back, between the particles is thus two seconds. For an expanding universe, this will gradually increase as time moves on. The **scaling factor** (or **expansion factor**), a, is the distance between the particles at any given time. Thus $a_0 = 1$, where a_0 is the value of a at the present time, and, for an expanding universe,

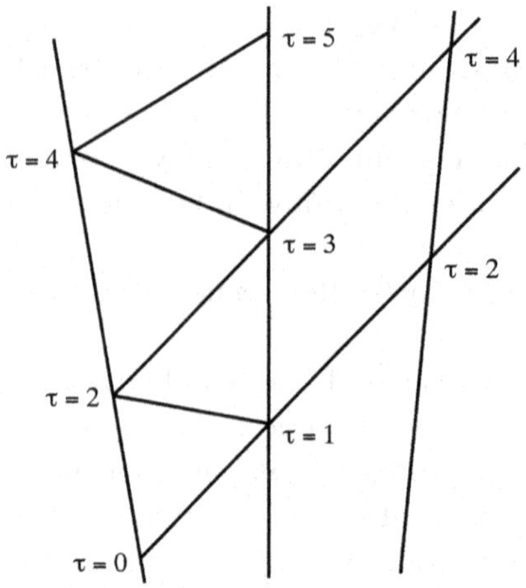

Figure 15.1. Values of the function τ. The world lines represent particles that have distance one light-second at the present time.

$\dot{a} > 0$. Note that, while a depends on what we take for our standard distance (or time) and is thus only defined up to a multiplicative constant, $\dot{a}/a$ is unique.

We also define an *elastic* time function, τ, such that light *always* takes light two τ-seconds to make the return journey between the particles (see Fig. 15.1), and thus

$$\Delta t = a\,\Delta\tau. \tag{15.1}$$

As time increases, a τ-second will gradually become longer and a clock reading τ-time will appear to run faster than a normal clock.

Consider, now, two *widely* separated, comoving particles. If a pulse of light with a τ-duration $\Delta\tau$ is sent out by the first particle and received with a τ-duration $\Delta\tau_0$ by the second particle, then $\Delta\tau = \Delta\tau_0$. Using equation (15.1), this gives $a_0\,\Delta t = a\,\Delta t_0$ and hence

$$a\,\omega = a_0\omega_0, \tag{15.2}$$

where ω and ω_0 are the transmitted and received frequencies. For an expanding universe, $a_0 > a$, and hence $\omega_0 < \omega$. In other words, a photon suffers a red shift in travelling between the particles.

Writing $d = t_0 - t$ and making a Taylor expansion of equation (15.2), we get

$$z = \frac{\omega - \omega_0}{\omega} = \frac{\dot{a}_0}{a_0}d + \left(\frac{\dot{a}_0}{a_0}\right)^2\left(1 - \frac{\ddot{a}_0 a_0}{\dot{a}_0^2}\right)d^2 + \mathcal{O}(d^3) \tag{15.3}$$

$$= Hd + H^2\left(1 + \frac{\Omega}{4}\right)d^2 + \mathcal{O}(d^3), \tag{15.4}$$

where

$$H = \frac{\dot{a}_0}{a_0} \quad \text{and} \quad \Omega = -\frac{2\ddot{a}_0 a_0}{\dot{a}_0^2}. \tag{15.5}$$

Taking $t = t_0$ to be the present time, $d = t_0 - t$ is the distance of a nearby, comoving particle (galaxy), and, to first order in d, equation (15.4) gives Hubble's law, $z = Hd$. The second term in equation (15.4) represents the deviation from Hubble's law for distant galaxies, and hence observations of such deviations can be used to estimate the value of Ω, though in practice this is extremely difficult. As we shall see in the next section, it is the value of Ω that determines whether the universe is open or closed.

15.3 The Evolution of the Universe

Let us now consider the equations of motion governing the evolution of the universe. In order to give a good account of these, it is necessary to use the Einstein equation, particularly when considering the early universe where relativistic effects due to radiation dominate. However, in the present era such effects are very small and a perfectly adequate set of equations of motion can be obtained very simply from Newton's law of gravity.

As the universe evolves, the scaling factor a will increase and the matter density will decrease – assuming, of course, we have an expanding universe. A comoving sphere of radius a, centered on some particle O, always contains the same number of particles, and hence its total mass $M = \frac{4\pi a^3 \rho}{3}$ will remain constant. This gives the matter conservation equation,

$$3\frac{\dot{a}}{a}\rho + \dot{\rho} = 0, \tag{15.6}$$

or, equivalently,

$$a^3\rho = \text{const.} \tag{15.7}$$

By Newton's law, a particle on the surface of the sphere will have an inward acceleration $\ddot{a}$ toward, the center given by

$$\ddot{a} = -Ma^{-2} = \frac{4\pi}{3}\rho a,$$

which, on using the matter conservation equation, integrates to give

$$\dot{a}^2 - \frac{8\pi}{3}\rho a^2 = -k, \tag{15.8}$$

for some constant k. Since a is determined up to a multiplicative factor, we may, if we wish, scale it to make k equal to one, zero, or minus one.

Given that the universe is now expanding (i.e. $\dot{a}_0 > 0$), equations (15.6) and (15.7) imply that a decreases and ρ increases as we move back in time, and, furthermore, that there existed (at least formally) a time in the *finite* past when $a=0$ and $\rho=\infty$. This is called the **big bang** and is a necessary consequence of the cosmological principle and Newton's equation of gravity. As we shall soon see, Einstein's equation implies very much the same thing. From now on, we make t unique by choosing it to be zero at the big bang. Thus

$$\lim_{t\to 0^+} a = 0 \quad \text{and} \quad \lim_{t\to 0^+} \rho = \infty,$$

and, as can easily be checked,

$$a = \mathcal{O}(t^{2/3}). \tag{15.9}$$

However, since ρ is finite at all *physical* events, there exists no point $p \in M$ such that $t(p) = 0$. The singular point (or points) representing the big bang do *not* belong to M, and t is a strictly positive function on M.

It is easy to solve equations (15.8) and (15.7). For $k = 1$, we get

$$a = \tfrac{1}{2}C(1 - \cos\eta) \quad \text{and} \quad t = \tfrac{1}{2}C(\eta - \sin\eta), \tag{15.10}$$

where η is some parameter and $C = 8\pi a^3\rho/3$. Similarly, for $k = -1$,

$$a = \tfrac{1}{2}C(\cosh\eta - 1) \quad \text{and} \quad t = \tfrac{1}{2}C(\sinh\eta - \eta). \tag{15.11}$$

For $k = 0$, we get the particularly simple equation

$$a = \left(\frac{9C}{4}\right)^{1/3} t^{2/3}. \tag{15.12}$$

In all cases we have arbitrarily high densities in the past, if $\dot{a}_0 > 0$. For $k = 1$, the universe will begin to contract after an initial phase of expansion – we call this a **closed universe.** However, for $k = -1$ or $k = 0$ the universe will continue to expand forever – an **open universe.**

For $k = 0$ the universe hovers between a closed and an open state. Using equations (15.8) and (15.5), we see that this gives a critical density

$$\rho_c = \frac{3H^2}{8\pi} \tag{15.13}$$

at the present time, $t = t_0$. The universe will thus be either open or closed according to whether ρ_0/ρ_c is greater or less than one. Furthermore, a simple calculation shows that $\Omega = \rho_0/\rho_c$. There are thus essentially two ways of determining whether the universe is open or closed. The first, and most direct, way is to determine the Hubble constant (and hence the critical density of the universe) and the present density of the universe. The second is to measure deviations from Hubble's law and use equation (15.3) to find Ω.

15.4 Horizons

Since $a \to 0$ as $t \to 0^+$, one might naively suppose that all comoving particles were once in causal contact with each other. We shall now show that this is in fact false, and that there exist comoving particles that could never have been in causal contact before any given time, t_0 say.

Choosing the elastic time function τ to be zero at the big bang, equation (15.1) gives

$$\tau(t) = \int_0^t \frac{ds}{a(s)}. \tag{15.14}$$

Note that equation (15.9) implies that this integral converges. By construction of the function τ, all comoving particles have a constant τ-distance apart. We thus obtain the stretched spacetime picture illustrated in Fig. 15.2, where comoving world lines are represented by vertical lines and null rays by lines at 45°.

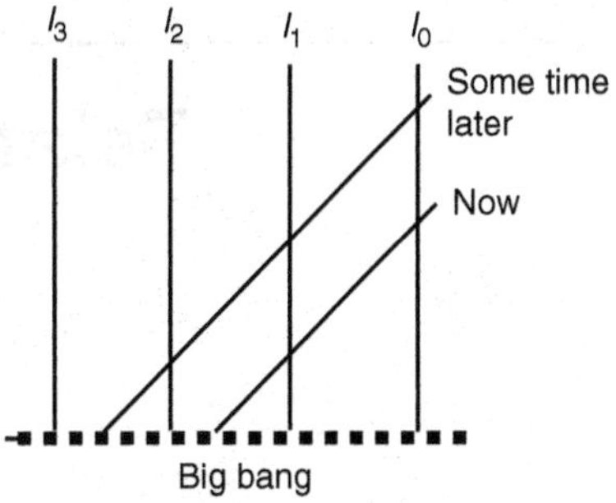

Figure 15.2. At the present time (now), the galaxy with world line l_1 is visible, but not the other two.

Referring to Fig. 15.2, let l_0 be the world line of our galaxy, and l_1, l_2, and l_3 the world lines of three distant galaxies. At the present time (now), the first galaxy will be visible, but not the other three – they still lie below the **horizon**. However, if we wait a while, the second galaxy will appear over the horizon, but the third will still be invisible. No event on world lines l_2 and l_3 can have any influence on our own galaxy at the present time – they are out of causal contact.

The **particle horizon** of a point P is defined to be P's past null cone (Fig. 15.3). Unlike the corresponding case in flat spacetime, the null cone is of *finite* extent, and there thus exist world lines that do not intersect it. A particle whose world line does not intersect P's horizon can have no causal influence on P.

Figure 15.3. The particle with world line l_1 can affect P, but not the particle with world line l_2.

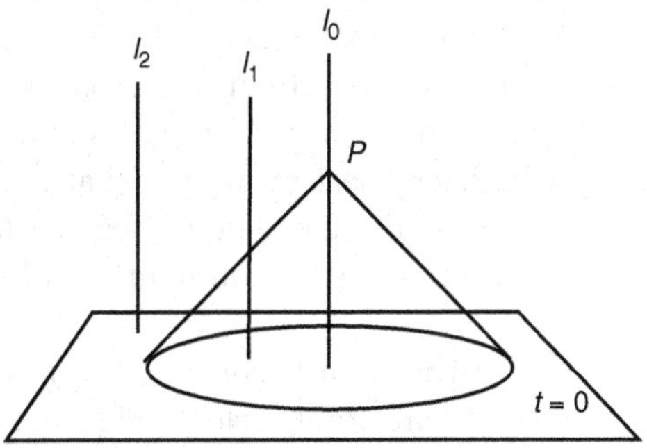

EXERCISES

15.1 If the universe were homogeneous but only isotropic about *one* point, would it be isotropic about all points?

15.2 Show that the red shift of a galaxy as it crosses the horizon (first becomes visible to us) is infinitely large.

15.3 If $a(t) = t^{2/3}$ and we see a galaxy cross the horizon at $t = t_0$, how long would we have to wait until its red shift decreased to 100%?

15.4 Show that $\Omega = \rho_0/\rho_c$.

15.5 If $k = 0$, find the age of the universe in terms of the present value of the Hubble constant H. If the universe were closed, would it be older or younger?

16

Relativistic Cosmology

Consider a spacetime M that, in addition to its metric g_{ab} and its various matter fields, contains a preferred function t (unique up to an additive constant) and a preferred four-velocity vector field v^a where $v^a \nabla_a t = 1$. The function t determines a family of hypersurfaces Σ_t on which t is constant, and v^a determines a timelike congruence with t as a parameter function. Our intention is to take M as model of the universe as a whole with the hypersurfaces Σ_t representing different eras, t representing universal time, and the curves of the congruence representing the world lines of comoving particles.

We say that a tensor field is **preferred** if it can be constructed from nothing more than the available structure on M, that is g_{ab}, t, v^a, and the various matter fields. For example, if a Maxwell field F_{ab} is one of the matter fields, then $E_a = F_{ab}v^b$ will be a preferred vector field. Using this notion, we impose the cosmological principle by demanding that the Σ_t surfaces be isotropic in that they contain no intrinsic, preferred vector fields and hence no preferred directions. This is a very strong condition and implies the following results:

(i) v^a is orthogonal to Σ_t. If not, then its projection in Σ_t would give a preferred vector field in contradiction to isotropy.

(ii) $v_a = \nabla_a t$. If $v_a - \nabla_a t \neq 0$, it would be orthogonal to v^a and hence a preferred vector field in Σ_t.

(iii) $Dv^a = 0$, that is, the world lines are geodesics. If $Dv^a \neq 0$, it would be orthogonal to v^a and therefore a preferred vector field in Σ_t.

(iv) Any preferred function is constant on Σ_t and hence depends only on t. If not, then its gradient would determine a preferred vector field in Σ_t.

(v) Any preferred, antisymmetric tensor F_{ab} must be zero. If not, F_{ab} would determine another antisymmetric tensor ${}^*F_{ab} = \frac{1}{2}\varepsilon_{abcd}F^{cd}$ and hence two vector fields, $E_a = F_{ab}v^b$ and $B_a = {}^*F_{ab}v^b$, orthogonal to v^a, neither of which can vanish if $F_{ab} \neq 0$.

(vi) Any preferred symmetric tensor k_{ab} must have the form $k_{ab} = \alpha v_a v_b + \beta h_{ab}$. Since k_{ab} is symmetric, it has four orthogonal eigenvectors. By isotropy, one of these must be v^a and the other two must have the same eigenvalue (β say); otherwise the one with the biggest eigenvalue would be a preferred vector field. Only tensors of the form $\alpha v_a v_b + \beta h_{ab}$ have this property.

(vii) Any preferred symmetric tensor k_{ab} satisfying $k_a{}^a = 0$ and $k_{ab}v^a = 0$ must be zero. This follows directly from the previous result.

(viii) The Weyl tensor C_{abcd} vanishes: C_{abcd} determines another tensor ${}^*C_{abcd} = \frac{1}{2}\varepsilon_{abef}C^{ef}{}_{cd}$ with the same properties as C_{abcd} and hence two symmetric vector fields, $E_{ab} = C_{adbc}v^d v^c$ and $B_{ab} = {}^*C_{adbc}v^d v^c$, with the properties described in the previous result. These must therefore vanish, which implies that the Weyl tensor vanishes.

From these results we see that all information about the curvature is contained in R_{ab}, that the Σ_t surfaces are spaces of **constant curvature**, that T_{ab} has the form of a **perfect fluid**, and that the shear s_{ab} of the congruence vanishes. The Raychaudhuri equation together with Einstein's field equation therefore gives

$$\dot{\theta} = \tfrac{1}{3}\theta^2 + 4\pi(3p + \rho) \tag{16.1}$$

and, since $\theta = -\nabla_a v^a$, the continuity equation (10.13) gives

$$\dot{\rho} - (\rho + p)\theta = 0. \tag{16.2}$$

We recall that $\theta = -\dot{V}/V$, where V is a small comoving volume of space (see equation 10.18). If we take V to be a small cube of side a, we have $V = a^3$ and

$$\theta = -3\frac{\dot{a}}{a}. \tag{16.3}$$

The function a, which we call the **scaling factor**, thus serves as a potential for θ and is defined up to a multiplicative constant. In terms of a, equations (16.1) and (16.2) become

$$-3\frac{\ddot{a}}{a} = 4\pi(3p + \rho) \tag{16.4}$$

and

$$\dot{\rho} = -3\frac{\ddot{a}}{a}(\rho + p), \tag{16.5}$$

which integrate to give the **Friedmann equation**

$$\left(\frac{\dot{a}}{a}\right)^2 - \frac{8\pi}{3}\rho = -\frac{k}{a^2}, \tag{16.6}$$

where k is a constant. We normalize a so that $k = 1, \ 0$, or -1. It is gratifying to note that, in the pressure-free case, equations (16.5) and (16.6) are the same as (15.6) and (15.8) in the previous chapter, thus giving further confirmation that Einstein's field equation reduces to Newtonian gravity in the appropriate limit.

From the above equations we see that all quantities are determined by a, the scaling factor, and k:

$$8\pi\rho = 3\left(\frac{\dot{a}}{a}\right)^2 + 3\frac{k}{a^2}, \tag{16.7}$$

$$8\pi p = -2\frac{\ddot{a}}{a} - \left(\frac{\dot{a}}{a}\right)^2 - \frac{k}{a^2}, \tag{16.8}$$

$$R_{ab} = 3\frac{\ddot{a}}{a}v_a v_b + \left[\frac{\ddot{a}}{a} + 2\left(\frac{\dot{a}}{a}\right)^2 + 2\frac{k}{a^2}\right]h_{ab}, \tag{16.9}$$

$$\nabla_a v_b = \frac{\dot{a}}{a}h_{ab}, \tag{16.10}$$

where $g_{ab} = v_a v_b + h_{ab}$. The last two equations follow from (10.10) and (10.19). In the case of a static universe where $\dot{a} = 0$, equation (16.9) reduces to

$$R_{ab} = 2\frac{k}{a^2}h_{ab}, \tag{16.11}$$

which shows that ka^{-2} is the curvature of the Σ_t surfaces [see equation (9.39)]. Even in the nonstatic case this still holds, because the *intrinsic* curvature of Σ_t depends only on a and not on $\dot{a}$. The function $\dot{a}$ determines the extrinsic curvature of Σ_t via equation (16.10)

In terms of the conformally rescaled metric

$$\hat{g}_{ab} = a^{-2}g_{ab} \tag{16.12}$$

the distance between two nearby, comoving particles will remain constant and the corresponding divergence function $\hat{\theta}$ will be zero. If $\hat{v}^a$ is the corresponding four-velocity vector field such that $\hat{g}_{ab}\hat{v}^a\hat{v}^b = 1$, we have

$$\hat{v}^a = av^a. \tag{16.13}$$

The corresponding parameter function $\hat{t}$ satisfying $1 = \hat{v}(\hat{t}) = a\dot{\hat{t}}$ is thus given by

$$dt = a\,d\hat{t} \tag{16.14}$$

or

$$\hat{t} = \int a^{-1}\,dt. \tag{16.15}$$

Comparing this equation with (15.14), we see that $\hat{t}$ is our old friend the elastic time τ. We thus have the important and useful result that all cosmological spaces are conformally related to a static cosmological space

$(M, \hat{g}_{ab}, \hat{v}^a, \hat{t})$ where $\hat{g}_{ab}$, $\hat{v}^a$, and $\hat{t}$ are given by equations (16.12), (16.13), and (16.15).

16.1 Friedmann Universes

A Friedmann universe is a cosmological spacetime where ρ and p are subject to the physically reasonable conditions

$$\rho \geq -3p, \qquad \rho \geq 0 \tag{16.16}$$

(i.e. the strong energy condition) together with some physically reasonable *equation of state* $\rho = F(p)$ relating ρ and p and depending on the type of matter we are considering. Note that, without an equation of state, the expansion factor a could be chosen arbitrarily, subject only to (16.16).

If we accept that the cosmological principle holds on a sufficiently large scale and that Einstein's field equation is a valid law of physics, we are led to the conclusion that some particular Friedmann universe will reflect the large-scale structure of the actual physical universe in which we live. All we have to do to determine this particular mathematical model of the physical universe is to find the equation of state for the large-scale distribution of matter in the universe, and then solve equations (16.4), (16.5), and (16.6) subject to the initial conditions corresponding to the present state of the universe.

The first striking result, which is independent of any particular equation of state and depends only on the dominant energy condition, follows directly from equation (16.4). According to this equation $\ddot{a} < 0$, and thus the universe cannot be static: at any given time it is either expanding ($\dot{a} > 0$) or contracting ($\dot{a} < 0$). Given that the universe is now in a state of expansion, $\ddot{a} < 0$ implies that the universe must have been expanding at a faster and faster rate as one goes backwards in time. There must therefore have been a time in the past when the distance between all "points in space" was zero and the density and curvature of spacetime was infinity. This singular state of the universe is referred to as the big bang. It is an inevitable consequence of the cosmological principle and Einstein's field equation.

The big bang does not represent an explosion of matter concentrated at a point of some *preexisting, nonsingular*, background spacetime, but a singularity in the structure spacetime itself. Since the spacetime structure is singular at the big bang, it does not make sense, either physically or mathematically, to ask about the state of the universe before the big bang; there is no natural way to extend the spacetime manifold and metric beyond the big-bang singularity.

For many years it was believed that the prediction of a singular origin of the universe was due merely to the assumption of exact homogeneity and isotropy, and if these assumptions were relaxed, one would get a nonsingular "bounce" at a small a rather than a singularity. However, as we have seen, the singularity theorems of general relativity show that singularities are generic features of cosmological solutions.

The two simplest equations of state are (dust) $p = 0$ and (radiation) $\rho = 3p$. For dust, equation (16.5) gives

$$\rho a^3 = \text{const.} \tag{16.17}$$

Note that this gives the lowest rate of decrease for ρ: if there is pressure, ρ will decrease more rapidly. At the opposite extreme, for radiation the same equation gives

$$\rho a^4 = \text{const.} \tag{16.18}$$

We thus see that although the radiation content of the universe in the present era may be negligible, its contribution to the total energy density far enough in the past should dominate over that of ordinary matter.

We have already considered solutions of the Friedmann equation for a dust-filled universe in the previous chapter. For a radiation-filled universe, the Friedmann equation becomes

$$\dot{a} - \frac{C}{a^2} + k = 0, \tag{16.19}$$

where $C = 8\pi\rho a^4/3$ is a constant. Solutions are given as follows:

$$a = \sqrt{C}\left[1 - \left(1 - \frac{t}{\sqrt{C}}\right)^2\right]^{1/2} \quad \text{for } k = 1, \tag{16.20}$$

$$a = (4C)^{1/4} t^{1/2} \quad \text{for } k = 0, \tag{16.21}$$

$$a = \sqrt{C}\left[\left(1 + \frac{t}{\sqrt{C}}\right)^2 - 1\right]^{1/2} \quad \text{for } k = -1. \tag{16.22}$$

Let us now consider the qualitative features of the evolution of the universe for any equation of state. If $k = 0$ or -1, equation (16.6) shows that $\dot{a}$ never can become zero. Thus, if the universe is presently expanding, it must expand forever. However, if $k = -1$, the universe cannot expand forever. Since ρ decreases more rapidly than a^{-2}, equation (16.6) implies that there exists a critical value, a_c, such that $a \le a_c$. Furthermore, a cannot asymptotically approach a_c as $t \to \infty$, because the magnitude of $\ddot{a}$ is bounded from below on account of equation (16.6). Thus, if $k = 1$, then at a finite time after the big bang the universe will achieve its maximum size a_c and thereafter contract.

16.2 The Cosmological Constant

As we have seen, there are no physically reasonable cosmological spaces that are static. When Einstein first applied his field equation to the cosmological problem, he very soon discovered this to be the case. Since at the time there was no observational evidence to suggest that the universe was in a nonstatic state and the philosophic prejudices of centuries underpinned the notion of a changeless universe, Einstein altered his field equation to include the cosmological constant. This had the effect of producing a repulsive force, which allowed, as we shall now show, a static cosmological space. If we include the cosmological constant in the picture, equations (16.4) and (16.6) become

$$-3\frac{\ddot{a}}{a} = 4\pi(3p + \rho) - \Lambda, \tag{16.23}$$

$$\left(\frac{\dot{a}}{a}\right)^2 - \frac{8\pi}{3}\rho - \frac{\Lambda}{3} = -\frac{k}{a^2}, \tag{16.24}$$

which have the static ($\dot{a} = 0$), pressure-free solution

$$4\pi\rho = \Lambda, \tag{16.25}$$

$$4\pi\rho = \frac{1}{a^2}, \tag{16.26}$$

where $k = 1$.

16.3 The Hot Big-Bang Model

Though the universe is not now in a state of thermal equilibrium, the blackbody signature of the background radiation strongly indicates that such a thermal state did exist in the past. The matter content of the universe now consists of ordinary matter – the stuff stars are made of – which is composed mostly of hydrogen atoms, the photons of the background radiation, a background of neutrinos (possibly of several different types), and possibly a considerable amount of **dark matter** of an unknown nature. Since hydrogen atoms are electrically neutral, they interact very weakly with photons and are thus not forced into a thermal state of the same temperature as the background radiation. Indeed, the formal notion of temperature is not applicable to the ordinary matter of the universe in the present era, because it is not in a thermal state – it forms a uniform dust distribution rather than a higgledy-piggledy thermal distribution. Whenever we refer to the temperature of the universe we shall mean the formal temperature of the background radiation. Though the neutrino background has yet to be observed, there are strong reasons to believe that this too has a blackbody thermal signature of a slightly lower temperature than the photon (microwave) background.

Roughly speaking, the temperature of the background radiation is a measure of the energy of an average photon. At the moment this energy is well below that necessary to ionize a hydrogen atom. However, as we shall see, the temperature is inversely proportional to the scaling factor and thus increases as we move back in time. There will therefore have been a time in the past when the temperature exceeded the ionization temperature, which is about 3000 K. Since the present temperature is about 3 K this will have occurred when the universe was 1000 times smaller than it is today, that is, when the distance between two comoving particles was 1000 times smaller than its present value. Before this epoch, hydrogen atoms would not have been able to exist, and the ordinary matter of the universe would have been in the form of free protons and electrons (and of course neutrons), which, being charged, interact strongly with photons. This presumably would have induced a state of thermal equilibrium, giving ordinary matter – now essentially in the form of free protons – a thermal distribution with the same temperature as the photon background.

The basic idea of the hot big-bang model of the universe is to assume that (global) thermal equilibrium did prevail before this time – when the temperature was above 3000 K – and to use the methods of standard thermodynamics and statistical physics to make physical and testable predictions.

16.4 Blackbody Radiation

Consider a gas of photons in thermal equilibrium with its surroundings at temperature T. A well-known result from statistical physics is that the number of photons per unit volume in angular-frequency range ω to $\omega + d\omega$ is given by

$$n_\gamma(\omega)\, d\omega = \frac{\omega^2}{\pi^2(e^{\omega/T} - 1)}\, d\omega. \tag{16.27}$$

Here we are using natural units of temperature in terms of which the Boltzmann constant k_B is identically unity. In such units T has dimension of energy or, equivalently, angular frequency, and thus ω/T is dimensionless. Since the energy of a photon is equal to its angular frequency (recall that we are using natural units in terms of which $\hbar = 1$), the energy per unit volume in frequency range ω to $\omega + d\omega$ is given by

$$\rho_\gamma(\omega)\, d\omega = \omega n(\omega)_\gamma\, d\omega = \frac{\omega^3}{\pi^2(e^{\omega/T} - 1)}\, d\omega. \tag{16.28}$$

The total energy per unit volume is thus given by

$$\rho_\gamma = \int_0^\infty \frac{\omega^3}{\pi^2(e^{\omega/T} - 1)}\, d\omega = \frac{\pi^2}{15} T^4 = \sigma T^4. \tag{16.29}$$

This is known as the Stefan–Boltzmann law, and $\sigma = \pi^2/15$, is sometimes called the Stefan–Boltzmann's constant. Similarly, the total number of photons per unit volume is given by

$$n_\gamma = \int_0^\infty n(\omega)_\gamma \, d\omega = \frac{T^3}{\pi^2} \int_0^\infty \frac{x^2}{e^x - 1} \, dx.$$

Since

$$\int_0^\infty \frac{x^2}{e^x - 1} \, dx = 2\zeta(3),$$

where ζ is the Riemann zeta function, this gives

$$n_\gamma = 2\frac{\zeta(3)}{\pi^2} T^3 \approx \frac{2.404}{\pi^2} T^3. \tag{16.30}$$

Furthermore, the entropy per unit volume of a photon gas is given by

$$s_\gamma = \tfrac{4}{3}\sigma T^3 = \tfrac{4}{45}\pi^2 T^3 = 3.6 n_\gamma. \tag{16.31}$$

We know that the background radiation has a blackbody spectrum in the present epoch, but will this be preserved as the universe evolves? We shall now show that this is indeed the case.

For a static model universe with metric $\hat{g}$ this will be trivially true, and we shall have

$$\hat{n}(\hat{\omega}) = \frac{\hat{\omega}^2}{e^{\hat{\omega}/\hat{T}} - 1}$$

where $\hat{\omega}$ and $\hat{T}$ are constant. We know that any model universe with metric g and expansion factor a is related to a static universe by $\hat{g}_{ab} = a^{-2} g_{ab}$. A small box of volume $\hat{V}$ with respect to $\hat{g}_{ab}$ will thus have volume $V = a^3 \hat{V}$ with respect to g_{ab}. Similarly, a photon with frequency $\hat{\omega}$ with respect to $\hat{g}_{ab}$ will have frequency $\omega = a^{-1}\hat{\omega}$ with respect to g_{ab}. The total number of photons with frequency between $\hat{\omega}$ and $\hat{\omega} + d\hat{\omega}$ contained in $\hat{V}$ will be $[\hat{n}(\hat{\omega}) \, d\hat{\omega}]\hat{V}$, and this will be equal to $[n(\omega) \, d\omega] V$, where $n(\omega)$ is the corresponding distribution with respect to g_{ab}. We thus have

$$[\hat{n}(\hat{\omega}) \, d\hat{\omega}]\hat{V} = [n(\omega) \, d\omega] V,$$

which gives

$$n(\omega) = a^{-2}\hat{n}(a\omega) = \frac{\omega^2}{e^{a\omega/\hat{T}} - 1}$$

and hence

$$n(\omega) = \frac{\omega^2}{e^{\omega/T} - 1}, \tag{16.32}$$

where

$$T = a^{-1}\hat{T}. \tag{16.33}$$

We thus see that the blackbody spectrum is preserved under the evolution of the universe, but the temperature is inversely proportional to the expansion factor and therefore increases as we move back in time. At the present time the temperature of the background radiation is quite low, but as we go back in time it increases, and near the big bang it can become arbitrarily high.

From equations (16.33), (16.29), and (16.31) we have

$$Ta = \text{const}, \tag{16.34}$$
$$\rho_\gamma a^4 = \text{const}, \tag{16.35}$$
$$s_\gamma a^3 = \text{const}. \tag{16.36}$$

Assuming that the nonradiative matter of the universe, consisting of massive particles, is adequately described by a dust distribution, its pressure p_M will be zero. This will certainly be true to a high degree of approximation for a thermal distribution at a temperature well below the threshold temperature $T_e = m_e$ of an electron – the lightest massive elementary particle. Under this approximation, the energy density of matter is $\rho = \rho_\gamma + \rho_M$, and its pressure is $p = p_\gamma = \rho_\gamma/3$. Putting all this into the conservation equation

$$\dot{\rho} = -3\frac{\dot{a}}{a}(\rho + p),$$

we get

$$\rho_M a^3 = \text{const}. \tag{16.37}$$

At the present time matter dominates over radiation ($\rho_\gamma \ll \rho_M$), and observations indicate that $\rho_\gamma/\rho_M \approx 10^{-3}$. However, as ρ_γ is proportional to a^{-4} while ρ_M is proportional to a^{-3}, this has not always been the case. When the universe was a thousand times smaller ($a_0/a = 1000$) and the temperature a thousand time higher than it is at present ($T = 3000$ K), ρ_γ would have been approximately equal to ρ_M and at all earlier times radiation would have dominated over matter. As we shall see in the next section, this occurred at about the same time when matter first became in full thermal contact with radiation. Between this time and the big bang we can thus use the methods of statistical mechanics to help us explain the evolution of the universe, and since radiation now dominates over matter, we can assume $\rho = \rho_\gamma$ and $p = \rho_\gamma/3$. Furthermore, since, for the early universe, ρ will dominate over the ka^{-2} term in the Friedmann equation, we may assume $k = 0$. Under these conditions, we have [see equation (16.21)]

$$a = (4C)^{1/4} t^{1/2}, \tag{16.38}$$

where $C = 8\pi\rho a^4/3 = \text{const}$. Since $Ta = \text{const}$, this implies that the temperature of the early universe is proportional to $t^{-1/2}$.

16.5 The Origin of the Background Radiation

In the present epoch the background radiation interacts very weakly with the rest of the matter in the universe, which consists mostly of hydrogen atoms. This is essentially because a hydrogen atom is electrically neutral and is thus comparatively unaffected by an electromagnetic field. There will exist a few photons in the high-energy tail of the blackbody spectrum with sufficient energy to ionize a hydrogen atom, but such events will be very rare. The photons of the background radiation may thus be considered to form an isolated system in its own right, without thermal contact with the rest of the matter in the universe. It is true the background radiation has a definite temperature, but only in the formal sense in that it has a blackbody spectrum. It is in no way in thermal equilibrium with the rest of the matter in the universe, which, indeed, cannot be considered to be a thermodynamical system: in the present era the notion of temperature is simply not applicable to the rest of the matter in the universe.

This, however, cannot always have been the case. As we have seen, the temperature of the background radiation becomes arbitrarily large as we move backwards in time, and we thus eventually reach a time when the majority of photons have an energy sufficient to ionize hydrogen atoms. Roughly speaking, this will occur when the temperature reaches the ionization energy i_H of a hydrogen atom (recall that we are using natural units in which temperature has the dimensions of energy), which corresponds to about 3000 K. In such an era, photons, electrons, and nuclei will form a plasma in thermal equilibrium, which will give rise to the blackbody spectrum of the background radiation. After this era, when the temperature drops below i_H, electrons and protons will combine into hydrogen atoms and will cease to interact with the photons, which, as we have seen, will retain their blackbody spectrum.

16.6 A Model Universe

As the purpose of this book is to give a description of the general structure of spacetime rather than a detailed account of our present knowledge of the universe, in this section we shall present a simple mathematical model of the universe containing only photons and baryons (protons and neutrons). We denote the energy densities and pressures of these two types of matter by ρ_γ, p_γ, ρ_b, p_b. There are, of course, other types of particle (electrons, neutrinos, gravitons, etc.) that contribute to the energy density of the universe and hence affect its evolution, but these either make a negligible contribution or else can be incorporated into the model once the general principles are grasped. There are also reasons to believe that much of the matter in the universe may be of a nonbaryonic nature (dark matter), but, for the sake of simplicity, we shall ignore this problem. In spite of this, our model does give quite a good explanation and description of the gross features of the universe. As with Dr. Johnson's dog on his hind legs, it is not done well, but you are surprised to find it done at all.

Let us assume that the photons form a thermal distribution, which, in the present era, describes the microwave background radiation, and that the baryons form a uniform dust distribution, which describes the remaining "hard matter" of the universe. In other words, we assume that $\rho_\gamma = 3p_\gamma$ and $p_b = 0$. This is a reasonable assumption to make about the present state of the universe, but not for earlier (hotter and denser) times when, presumably, both forms of matter where in thermal equilibrium. At such times the thermal motion of baryons would cause p_b to be nonzero. However, in the actual universe in which we live, the ratio of the number of photons to baryons in any comoving volume of space is of the order of 10^9. In spite of this enormous ratio, the energy of an average photon is so small at present that we have $\rho_b \gg \rho_\gamma$. However, as we have seen, the energy of a photon increases as we move back in time, and for the early universe there would have been 10^9 *high-energy* photons to every baryon. In other words, the microwave background radiation – which is now no more than a barely perceivable sprinkling of photons in a universe consisting of baryonic matter (the stuff of all ordinary matter) – once dominated the universe. At sufficiently early times ordinary matter (the stuff you and I and galaxies are made of) would have been no more than a very slight seasoning in a great primeval stew of photons. Thus, though both forms of matter were once in thermal equilibrium, the baryonic pressure would have made a negligible contribution at such times.

Conservation of energy implies

$$\dot{\rho}_b = -3\frac{\dot{a}}{a}\rho_b, \tag{16.39}$$

$$\dot{\rho}_\gamma = -3\frac{\dot{a}}{a}(\rho_\gamma + p_\gamma) = -4\frac{\dot{a}}{a}\rho_\gamma, \tag{16.40}$$

and hence $a^3\rho_b$ and $a^4\rho_\gamma$ are constant. Furthermore, the equation

$$3\frac{\ddot{a}}{a} = -4\pi(\rho + 3p)$$

[see equation (16.4)] implies that the quantity

$$\dot{a}^2 - \frac{8\pi}{3}(\rho_\gamma + \rho_b)a^2 = -k \tag{16.41}$$

is constant. As we have seen, the curvature scalar of the $t = \text{const}$ sections is given by

$$K = \frac{k}{a^2}. \tag{16.42}$$

We recall that the scaling factor a is the distance between two arbitrary neighboring comoving particles and is thus defined up to a multiplicative factor. If $K \neq 0$, the standard way of fixing this factor is to choose k to be either 1 (a closed universe) or -1 (an open universe), in which case $a = R$ where $R^2 = |K|^{-1}$. The function R, which may be interpreted as the "radius" of the universe, provides a convenient standard, comoving measure of distance. However, if $K = 0$ – and there are indications that this may be so for the actual universe – we cannot normalize a in this way. For this reason we shall choose a nonstandard normalization for a such that a region of volume $4\pi a^3/3$ contains, on the average, just one baryon. In terms of this normalization,

$$\frac{4\pi}{3} a^3 \rho_b = m_b, \tag{16.43}$$

where m_b is the mass of one proton. With this choice of scaling factor, $|k| \neq 1$ but is given by

$$|k| = \left(\frac{a}{R}\right)^2. \tag{16.44}$$

At the present time the baryon density of the universe is very roughly one baryon to every cubic yard, and hence a_0 (the present value of a) is about one yard. On the other hand, R is either infinite or equal to some huge distance, and hence we may safely assume that k is some very small number. Fortunately, if we restrict attention to the early universe, perhaps even up to the present epoch, the constant k plays no significant role at all.

A much more important constant, at least as far as the early universe is concerned, is given by

$$C = \frac{8\pi}{3} \rho_\gamma a^4. \tag{16.45}$$

As we shall see, this constant is closely related to the magic ratio, 10^9, of photons to baryons.

In terms of the constants C, m_b, and k, equation (16.41) becomes

$$\dot{a}^2 = \frac{C}{a^2} + 2\frac{m_b}{a} - k, \tag{16.46}$$

or, by taking t to be a function of a,

$$\frac{dt}{da} = a(C + 2m_b a - ka^2)^{-1/2}. \tag{16.47}$$

Note that for small a this becomes

$$\frac{dt}{da} = C^{-1/2} a - m_b C^{-3/2} a^2, \tag{16.48}$$

which shows that k plays no significant role in the early universe.

Equations (16.45) and (16.43) give

$$\rho_\gamma = \frac{3}{8\pi} C a^{-4} \quad \text{and} \quad \rho_b = \frac{3}{4\pi} m_b a^{-3}, \tag{16.49}$$

which, together with the solution of equation (16.47), determine ρ_γ and ρ_b as functions of t – assuming, of course, that we know the value of C. In terms of temperature, ρ_γ is given by

$$\rho_\gamma = \frac{\pi^2}{15} T^4 \tag{16.50}$$

[see (16.29)] which, together with equation (16.49), gives

$$\frac{45}{8\pi^3} C = (Ta)^4. \tag{16.51}$$

This gives T as a function of a, and hence [given the solution of (16.47)] as a function of t. Again, this depends of knowing the value of the mysterious constant C.

In order to get a handle on this constant, we introduce the photon entropy density, which is given by

$$s_\gamma = \tfrac{4}{45}\pi^2 T^3 \approx 3.6 n_\gamma \tag{16.52}$$

[see (16.31)]. The photon entropy per baryon (specific entropy) is thus given by

$$S = s_\gamma \left(\frac{4\pi}{3} a^3 \right) = \frac{16\pi^3}{135} (Ta)^3, \tag{16.53}$$

which together with (16.51) gives

$$\left(\frac{S}{2} \right)^4 = \frac{1}{5} \left(\frac{2\pi C}{3} \right)^3. \tag{16.54}$$

We thus have a relation between C and S. Since, by (16.52), $s_\gamma \approx 3.6 n_\gamma$, where n_γ is the photon number density, we see that the ratio of photons to baryons – that is, the number of photons contained in a volume $4\pi a^3/3$ – is approximately equal to $0.28S$. Taking this ratio to be about 10^9, corresponding to the current physical estimate, an approximate value of C can be found using (16.54). With this value of C, we can now determine the values of the physical quantities ρ_γ, ρ_b, and T at any instant of universal time t, and thus obtain the evolution of the gross features of the universe – at least according to our simple model. It must, however, be emphasized that all quantities are expressed in natural units where $c = \hbar = G = k_B = 1$.

EXERCISES

16.1 Given a three-space of constant curvature K, there exists, locally, a coordinate system (r, θ, φ) such that the metric has the form

$$h = \frac{dr^2}{1 - Kr^2} + r^2(d\theta^2 + \sin^2\theta\, d\varphi^2).$$

Use this fact to show that the metric of a comological spacetime can be written locally as

$$g = dt^2 - \frac{a^2\, dr^2}{1 - kr^2} + r^2(d\theta^2 + \sin^2\theta\, d\varphi^2).$$

Can this also be taken to be a global result?

16.2 Verify equations (16.20)–(16.22).

16.3 A null cone with vertex at $t = t_0$ intersects Σ_t in a two-sphere. Find an expression for the area of this two-sphere for a closed universe with $k = 1$.

16.4 For a radiative universe ($\rho = 3p$) with $k = 0$ show that

$$T^4 = \frac{3t^2}{32a\pi}.$$

16.5 The energy density ρ of a photon gas in thermal equilibrium depends only on its temperature and satisfies the equation of state $\rho = 3p$. Use this fact together with the thermodynamical relation

$$T\, dS = dE + p\, dV$$

(T = temperature, S = entropy, and V = volume) to show that

$$\rho = \sigma T^4 \quad \text{and} \quad s = \tfrac{4}{3}\sigma T^3$$

where σ is some constant. Give an argument that the total entropy of a comoving volume remains constant, and show that $Ta = \text{const}$.

16.6 Show that the total life span (big bang to big crunch) of a closed, radiative universe with $k = 1$ is given by $\Delta t = 2\sqrt{C}$, where $C = 8\pi\rho a^4/3$. (Note that C is a constant for a radiative universe.)

16.7 An observer lives in the type of universe described in the previous exercise. He looks through a very powerful telescope and sees the back of his own head. What can he say about the value of C and the topology of his universe?

Solutions and Hints to Selected Exercises

Chapter 2

2.3 S is the intersection of $N^+(p)$ and $N^+(q)$. Since $\dim N^+(p) = \dim N^+(q) = 3$ and $\dim M = 4$, we have $\dim S = 2$ if S is nonempty. If p occurs before q to some observer, then $N^+(p)$ lies inside $N^+(q)$ and S is empty.

Chapter 3

3.1 Referring to Fig. 3.10, we have $t_2' = K^{-1}t_2$ and $t_1' = Kt_1'$, where

$$K = \sqrt{\frac{1+v}{1-v}}.$$

Thus

$$x' = \frac{K^{-1}t_2 - Kt_1}{2} = \frac{K^{-1}(t+x) - K(t-x)}{2}$$
$$= \frac{x - vt}{\sqrt{1-v^2}}.$$

Similarly,

$$t' = \frac{t - vx}{\sqrt{1-v^2}}.$$

3.2 Both Fig. S.1(a) and (b) lead to

$$K = \sqrt{\frac{1+v}{1-v}}.$$

In the case of (b), where the particles are moving towards each other, the Doppler shift is given by

$$f_{\text{re}} = Kf_{\text{tr}} = f_{\text{tr}}\sqrt{\frac{1+v}{1-v}}.$$

3.3 From Fig. S.2 we see that Peter will be $K + K^{-1}$ years old on Pat's return. Since $v = \frac{1}{2}$, this give

$$K + K^{-1} = \sqrt{3} + \frac{1}{\sqrt{3}}.$$

3.4 Figure S.3 illustrates the situation where the events "guard enters tunnel" and "driver leaves tunnel" occur at the same time according to a "stationary" observer in the tunnel. t is the proper time with respect to the tunnel, and t' is the proper time with respect to the train. The length of the tunnel is l. From the diagram, we see that the length of the train is given by $l(K + K^{-1})/2$. Since $v = \sqrt{3}/2$ and

$$K = \sqrt{\frac{1+v}{1-v}},$$

we see that $l(K + K^{-1})/2 = 2l$.

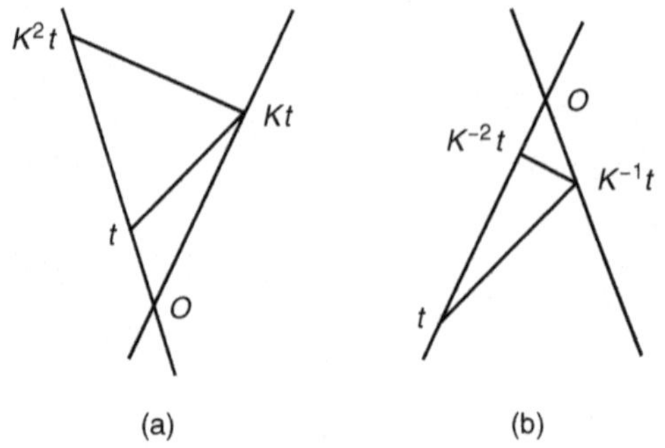

Figure S.1.

3.5 In general, the angle will depend on the state of motion of the observer. For the complete solution of this problem, see Example 4.1.

Chapter 4

4.1 Since $\boldsymbol{p}$ is null and $\boldsymbol{p} \cdot \boldsymbol{e}_1 = \boldsymbol{p} \cdot \boldsymbol{e}_2 = 0$, we may write $\boldsymbol{p} = \boldsymbol{e}_0 + \boldsymbol{e}_3$. The current density vector will have the form $\boldsymbol{j} = \rho\boldsymbol{p} = \rho(\boldsymbol{e}_0 + \boldsymbol{e}_3)$. Since this implies that $\rho = \boldsymbol{j} \cdot \boldsymbol{v} = \rho(\boldsymbol{e}_0 + \boldsymbol{e}_3) \cdot \boldsymbol{e}_0$, ρ is the particle density according to an observer sitting at the north pole. An observer moving vertically upwards with speed W will have four-velocity $\boldsymbol{v}' = \gamma(\boldsymbol{e}_0 + W\boldsymbol{e}_3)$, where

$$\gamma = \frac{1}{\sqrt{1 - W^2}}.$$

Thus the corresponding density is given by

$$\rho' = \boldsymbol{j} \cdot \boldsymbol{v}' = \gamma\rho(\boldsymbol{e}_0 + \boldsymbol{e}_3) \cdot (\boldsymbol{e}_0 + W\boldsymbol{e}_3) = \gamma\rho(1 - W).$$

An observer moving horizontally in direction $\boldsymbol{e}_1$ with speed W will have four-velocity $\boldsymbol{v}' = \gamma(\boldsymbol{e}_0 + W\boldsymbol{e}_1)$. Thus the corresponding density is given by

$$\rho' = \boldsymbol{j} \cdot \boldsymbol{v}' = \gamma\rho(\boldsymbol{e}_0 + \boldsymbol{e}_3) \cdot (\boldsymbol{e}_0 + W\boldsymbol{e}_1) = \gamma\rho.$$

Figure S.2.

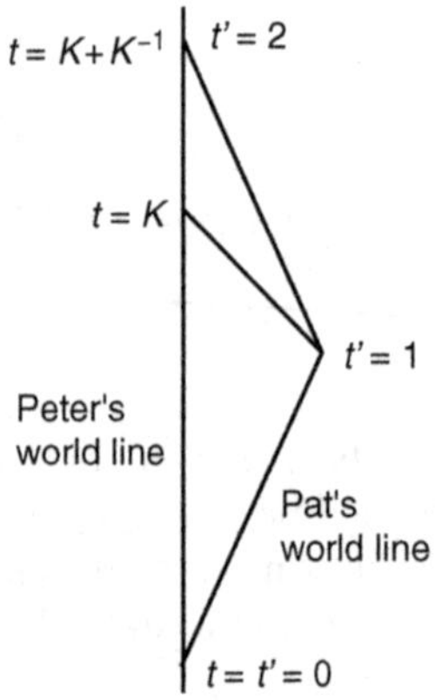

4.2 Let $\boldsymbol{v}$ and $\boldsymbol{v}'$ be the four-velocities of the two observers, and $\boldsymbol{a}$ the connecting vector such that $\boldsymbol{v} \cdot \boldsymbol{a} = \boldsymbol{v}' \cdot \boldsymbol{a} = 0$. This condition implies that a given by $a^2 = -\boldsymbol{a} \cdot \boldsymbol{a}$ is the minimum distance between the two observers. Choosing O (see Fig. S.4) as an origin point of position vector $\boldsymbol{x}$, the two world lines may be expressed as

$$\boldsymbol{x} = \boldsymbol{v}t \quad \text{and} \quad \boldsymbol{x} = \boldsymbol{v}'t' + \boldsymbol{a}, \tag{S.1}$$

where t and t' represent proper time on the respective world lines. Given a point P with proper time t on the first world line, there will exist a corresponding point Q with proper time t' on the second world line such that $\boldsymbol{n} = QP$ is null. A photon sent from Q will thus be received at P. Since

$$\boldsymbol{n} = \boldsymbol{v}t - \boldsymbol{v}'t' - \boldsymbol{a},$$

the null condition $\boldsymbol{n} \cdot \boldsymbol{n} = 0$ gives

$$t^2 + t'^2 - 2\gamma tt' - a^2 = 0,$$

where

$$\gamma = \boldsymbol{v} \cdot \boldsymbol{v}' = \frac{1}{\sqrt{1 - v^2}},$$

and v is the relative speed between the two observers. Solving for t' in terms of t and choosing the solution such that Q lies to the past of P gives

$$t' = \gamma t - \sqrt{a^2 + \gamma^2 v^2 t^2} \tag{S.2}$$

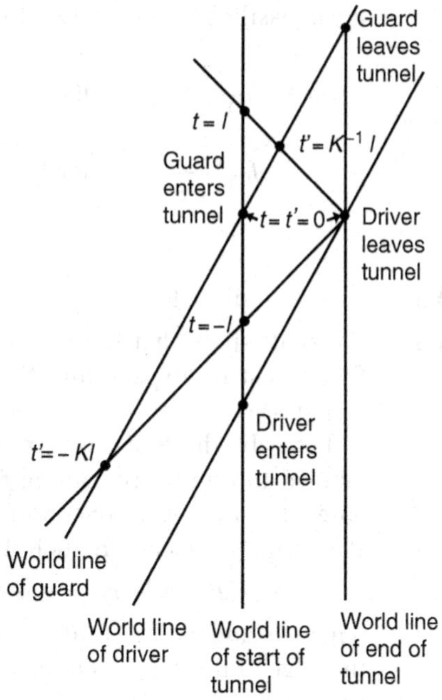

Figure S.3.

and hence

$$\boldsymbol{n} = \boldsymbol{v}t - \boldsymbol{v}'(\gamma t - \sqrt{a^2 + \gamma^2 v^2 t^2}) - \boldsymbol{a}. \tag{S.3}$$

Let $\hat{\boldsymbol{n}} = \alpha \boldsymbol{n}$ and choose α such that $\hat{\boldsymbol{n}} \cdot \boldsymbol{v}' = 1$. This gives

$$\alpha = \frac{1}{\sqrt{a^2 + \gamma^2 v^2 t^2}} \tag{S.4}$$

and hence

$$\hat{\boldsymbol{n}} \cdot \boldsymbol{v} = \gamma \frac{\sqrt{a^2 + \gamma^2 v^2 t^2} - \gamma v^2 t}{\sqrt{a^2 + \gamma^2 v^2 t^2}}. \tag{S.5}$$

Taking $\hat{\boldsymbol{n}}$ to be the four-momentum of a photon traveling between Q and P, the condition $\hat{\boldsymbol{n}} \cdot \boldsymbol{v}' = 1$ says that the photon has unit frequency according to the observer at Q. The photon's frequency as observed at P will thus be given by equation (S.5). The required Doppler equation is therefore given by

$$f_{\text{re}} = f_{\text{tr}} \gamma \frac{\sqrt{a^2 + \gamma^2 v^2 t^2} - \gamma v^2 t}{\sqrt{a^2 + \gamma^2 v^2 t^2}}. \tag{S.6}$$

Figure S.4.

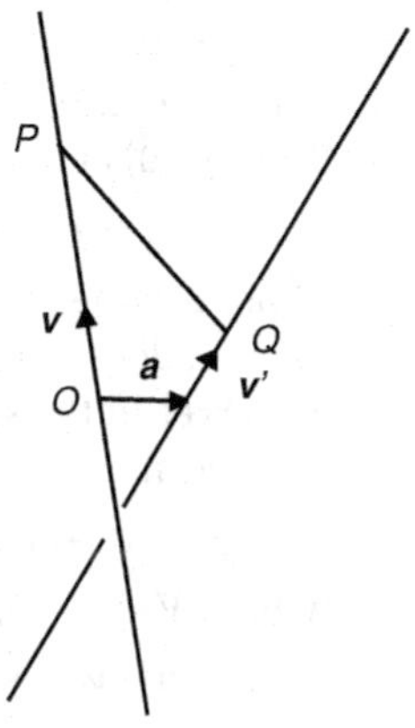

It can easily be checked that this gives the known equations

$$f_{\rm re} = f_{\rm tr}\sqrt{\frac{1-v}{1+v}} \qquad \text{for } t > 0, \tag{S.7}$$

$$f_{\rm re} = f_{\rm tr}\sqrt{\frac{1+v}{1-v}} \qquad \text{for } t < 0 \tag{S.8}$$

if $a = 0$.

4.3 Use Example 4.1.

4.4 We shall show that the moon still appears as a disk to an observer flying past in a spaceship, though the apparent center of the moon changes.

Let $\boldsymbol{v}$ be the four-velocity of an observer on earth, and let $\boldsymbol{n}_0$ be tangent to a null ray coming from the center of the moon and $\boldsymbol{n}$ be tangent to any null ray coming from the apparent edge of the moon. We normalize $\boldsymbol{n}_0$ and $\boldsymbol{n}$ such that $\boldsymbol{v}\cdot\boldsymbol{n}_0 = \boldsymbol{v}\cdot\boldsymbol{n} = 1$ and write

$$\boldsymbol{n} = \boldsymbol{v} + \boldsymbol{e} \quad \text{and} \quad \boldsymbol{n}_0 = \boldsymbol{v} + \boldsymbol{e}_0,$$

where $\boldsymbol{e}$ and $\boldsymbol{e}_0$ are unit vectors orthogonal to $\boldsymbol{v}\cdot\boldsymbol{e}$ points toward the edge of the moon, and $\boldsymbol{e}_0$ points towards its center. Orthogonality implies

$$\boldsymbol{n}\cdot\boldsymbol{n}_0 = 1 + \boldsymbol{e}\cdot\boldsymbol{e}_0 = C. \tag{S.9}$$

Since the moon appears as a disk, $\boldsymbol{e}\cdot\boldsymbol{e}_0$, and hence $C = \boldsymbol{n}\cdot\boldsymbol{n}_0$, will remain constant as $\boldsymbol{e}$ circles round the apparent edge of the moon.

Let $\boldsymbol{v}'$ be the four-velocity of the observer flying past in a spaceship, and let

$$\boldsymbol{n}' = \omega\boldsymbol{n} \quad \text{and} \quad \boldsymbol{n}'_0 = \omega_0\boldsymbol{n}_0,$$

where we choose ω and ω_0 such that $\boldsymbol{v}'\cdot\boldsymbol{n}'_0 = \boldsymbol{v}'\cdot\boldsymbol{n}' = 1$. Contracting the above equations with $\boldsymbol{v}'$, we find that

$$\boldsymbol{n}\cdot\boldsymbol{v}' = \omega^{-1} \quad \text{and} \quad \boldsymbol{n}_0\cdot\boldsymbol{v}' = \omega_0^{-1}.$$

Similarly, contracting with $\boldsymbol{v}$, we find that

$$\boldsymbol{n}'\cdot\boldsymbol{v} = \omega \quad \text{and} \quad \boldsymbol{n}'_0\cdot\boldsymbol{v} = \omega_0. \tag{S.10}$$

Writing

$$\boldsymbol{n}' = \boldsymbol{v}' + \boldsymbol{e}' \quad \text{and} \quad \boldsymbol{n}'_0 = \boldsymbol{v}' + \boldsymbol{e}'_0,$$

where $\boldsymbol{e}'$ and $\boldsymbol{e}'_0$ are unit vectors orthogonal to $\boldsymbol{v}'$, and using equations (S.9) and (S.10), we get

$$\boldsymbol{n}'\cdot\boldsymbol{n}'_0 = 1 + \boldsymbol{e}'\cdot\boldsymbol{e}'_0 = C\omega\omega_0. \tag{S.11}$$

Since ω varies as $\boldsymbol{n}'$ circles the edge of the moon, this equation implies that $\boldsymbol{n}'_0$ does not point toward the center of the moon according to the observer in the spaceship.

Writing

$$\boldsymbol{v} = \gamma(\boldsymbol{v}' + \boldsymbol{V}),$$

where $\boldsymbol{V}\cdot\boldsymbol{v}' = 0$, equation (S.10) gives

$$\omega = \boldsymbol{n}'\cdot\boldsymbol{v} = \gamma(\boldsymbol{v}' + \boldsymbol{e}')\cdot(\boldsymbol{v}' + \boldsymbol{V}) = \gamma(1 + \boldsymbol{e}'\cdot\boldsymbol{V}).$$

Substituting this into equation (S.11), we get

$$e' \cdot e'_0 = C\omega_0\gamma(1 + e' \cdot V) - 1,$$

which gives

$$e' \cdot a = C\omega_0\gamma - 1, \qquad \text{(S.12)}$$

where $a = e'_0 + C\omega_0\gamma V$. Since $C\omega_0\gamma - 1$ and a are constant, this shows that the moon still appears as a disk to the observer in the spaceship, but that a, rather than e'_0, points toward the apparent center of the moon.

Chapter 5

5.1 The four-momentum of electron+positron is timelike, but the four-momentum of a single photon is null. Conservation of four-momentum thus implies that electron+positron cannot decay into a single photon.

5.2 The four-momentum of the particle is Mv, where v is its four-velocity. The two photons will have four-momenta of the form $p_1 = E(v + e)$ and $p_2 = E(v - e)$, where e is a unit vector orthogonal to v. Conservation gives $Mv = 2Ev$ and hence $E = M/2$. The energy of each of the photons is given by $p_1 \cdot v = E = M/2$.

5.3 Let p be the four-momentum of the photon, and mw the four-momentum of the particle of mass m. Conservation gives

$$p = Mv - mw.$$

Contracting with v gives $E = p \cdot v = M - m\gamma$, where $\gamma = v \cdot w$. Furthermore, since p is null, we have

$$p \cdot p = M^2 - 2\gamma mM + m^2 = 0.$$

Thus

$$\gamma = \frac{M^2 + m^2}{2Mm}$$

and hence

$$E = M - \frac{M^2 + m^2}{2M}.$$

5.4 Conservation gives

$$m_1 v_1 = Mv - m_2 v_2$$

and hence

$$m_1^2 = M^2 - 2Mm_2\gamma_2 + m_2^2,$$

where $g_2 = v \cdot v_2$. Solving for γ_2, we get

$$\gamma_2 = \frac{M^2 + m_2^2 - m_1^2}{2Mm_2}.$$

The energy of particle 2 is given by $E_2 = m_2 v_2 \cdot v = m_2\gamma_2$, and hence

$$E_2 = \frac{M^2 + m_2^2 - m_1^2}{2M}.$$

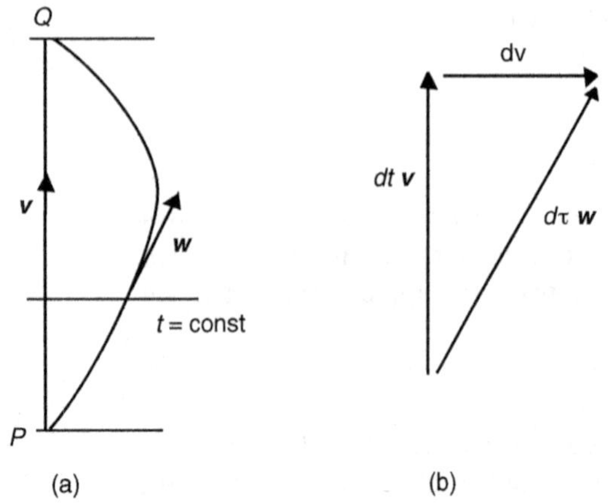

Figure S.5.

Similarly,

$$E_2 = \frac{M^2 + m_1^2 - m_2^2}{2M}.$$

5.6 Consider two world lines connecting points P and Q: a straight one with four-velocity $\boldsymbol{v}$ and proper time t, and a curved one with four-velocity tangent vector $\boldsymbol{w}$ and proper time τ [see Fig. S.5(a)]. From Fig. S.5(b), we see that

$$d\tau\, \boldsymbol{w} = dt\, \boldsymbol{v} + d\boldsymbol{V},$$

where $\boldsymbol{v} \cdot \partial \boldsymbol{V} = 0$ and dt and $d\tau$ are corresponding increments in t and τ. Contracting with $\boldsymbol{v}$, we get $\gamma\, d\tau = dt$, where $\gamma = \boldsymbol{v} \cdot \boldsymbol{w}$. Thus

$$\Delta\tau = \int_P^Q d\tau = \int_P^Q \gamma^{-1}\, dt,$$

$$\Delta t = \int_P^Q dt,$$

and therefore $\Delta\tau \leq \Delta t$, since $\gamma^{-1} \leq 1$. A straight world line between P and Q thus contains maximum proper time.

Chapter 6

6.2 Using the antisymmetry of ε_{abcd} together with equation (6.15), we have

$$\begin{aligned} l_a &= \varepsilon_{abcd}\varepsilon^{bcdh} j_h \\ &= -\varepsilon_{bcda}\varepsilon^{bcdh} j_h \\ &= 6\delta_a^h j_h \\ &= 6 j_a. \end{aligned}$$

6.3 Writing

$$4\pi T_{ab} = F_{ac}F_b{}^c - \tfrac{1}{4} g_{ab} F_{de}F^{de},$$

we have

$$\begin{aligned} 4\pi T_{ba} &= F_{bc}F_a{}^c - \tfrac{1}{4} g_{ab} F_{de}F^{de} \\ &= F_a{}^c F_{bc} - \tfrac{1}{4} g_{ab} F_{de}F^{de} \\ &= F_{ac}F_b{}^c - \tfrac{1}{4} g_{ab} F_{de}F^{de} \\ &= 4\pi T_{ab}. \end{aligned}$$

Furthermore, using the fact that $g_a^a = 4$, we have

$$\begin{aligned} 4\pi T_a^a &= F_{ac}F^{ac} - \tfrac{1}{4} g_a^a F_{de}F^{de} \\ &= F_{ac}F^{ac} - F_{de}F^{de} \\ &= 0. \end{aligned}$$

6.4 Use equation (6.17).

6.5 That

$$8\pi T_{ab} = -(F_{ac}F_b{}^c + {}^*F_{ac}{}^*F_b{}^c)$$

follows trivially from the previous exercise. Contracting with $v^a v^b$ and using the fact that $E_a = F_{ab}v^a$ and $B_a = {}^*F_{ab}v^a$, we get

$$8\pi T_{ab}v^a v^b = -(E_c E^c + B_c B^c).$$

This implies $T_{ab}v^a v^b > 0$, since E_a and B_a are spacelike and can't both vanish for $F_{ab} \neq 0$.

6.6 Consider the space W of all vectors of the form $w_a = F_{ab}v^b$. Since $F_{[ab}F_{c]d} = 0$, we have $F_{[ab}w_{c]} = 0$ for all $w_a \in W$. From this it can be seen that dim $W = 2$ and that F_{ab} is proportional to $U_{[a}V_{b]}$, where (U_a, V_a) is a basis for W.

Chapter 7

7.1 Let w^a be the four-velocity of a noncomoving observer. By Example 7.3 a Coulomb field is given by

$$F_{ab} = 2e\frac{v_{[a}e_{b]}}{r^2},$$

and its magnetic component with respect to w^a is

$$B_a = \tfrac{1}{2}\varepsilon_{abcd}w^b F^{cd}.$$

Thus,

$$B_a = er^{-2}\varepsilon_{abcd}w^b v^c e^d.$$

Writing

$$w^b = \gamma(v^b + V^b),$$

where $V^a v_a = 0$ and $\gamma = v_a w^a$, this gives

$$B_a = e\gamma r^{-2}\varepsilon_{abcd}V^b v^c e^d.$$

In three-vector notation, this becomes

$$\boldsymbol{B} = e\gamma r^{-2}\boldsymbol{V} \times \mathbf{e}.$$

7.2 For a Coulomb field, $A_a = er^{-1}v_a$, and for a plane wave (see Example 7.4) $A_a = s_a \cos(n_b x^b)$.

7.3 For a plane wave, $F_{ab} = s_{[a}n_{b]}\sin(n_c x^c)$. Therefore

$$\begin{aligned} E_a &= F_{ab}v^b = \tfrac{1}{2}(s_a n_b - s_b n_a)v^b \sin(n_c x^c) \\ &= \tfrac{1}{2}(s_a n_b v^b - s_b v^b n_a)\sin(n_c x^c), \\ B_a &= {}^*F_{ab}v^b = \tfrac{1}{2}\varepsilon_{abcd}v^b s^c n^d \sin(n_c x^c), \end{aligned}$$

and, by the antisymmetry of ε_{abcd}, we have $B_a E^a = 0$.

7.5

$$\begin{aligned} \mathcal{L}_v w_a &= \mathcal{L}_v(g_{ab}w^b) = w^b\mathcal{L}_v g_{ab} \\ &= w^b(g_{eb}\nabla_a v^e + g_{ae}\nabla_b v^e) \\ &= w^b(\nabla_a v_b + \nabla_b v_a). \end{aligned}$$

Thus $\mathcal{L}_v w^a = 0$ implies $\mathcal{L}_v w_a = 0$ only if $w^b(\nabla_a v_b + \nabla_b v_a) = 0$.

7.6
$$\mathcal{L}_v T_{ab}{}^c = v^e \nabla_e T_{ab}{}^c + T_{eb}{}^c \nabla_a v^e + T_{ae}{}^c \nabla_b v^e - T_{ab}{}^e \nabla_e v^c.$$

7.7
$$\nabla^{a*} F_{ab} = \nabla^a \varepsilon_{abcd} F^{cd}/2 = \varepsilon_{abcd} \nabla^a F^{cd}/2.$$

Thus $\nabla^{a*} F_{ab} = 0$ implies $\varepsilon_{abcd} \nabla^a F^{cd} = 0$, which implies $\nabla^{[a} F^{cd]} = 0$.

7.8 The conservation equation $\nabla^a T_{ab} = 0$, follows trivially from $\nabla^{a*} F_{ab} = \nabla^a F_{ab} = 0$ on writing T_{ab} in the form

$$8\pi T_{ab} = -(F_{ac} F_b{}^c + {}^*F_{ac}{}^*F_b{}^c)$$

(see Exercise 6.5).

7.9 This follows from Theorem 7.1.

Chapter 8

8.2 In Example 6.1 we showed that

$$F_{ab} = E_{ab} - {}^*B_{ab},$$
$${}^*F_{ab} = {}^*E_{ab} + B_{ab},$$

where $B_{ab} = 2B_{[a} v_{b]}$ and $E_{ab} = 2E_{[a} v_{b]}$. Writing the Maxwell equation $\nabla^a F_{ab} = -4\pi J_b$ in terms of E_{ab}, ${}^*B_{ab}$, ρ_0, and K_a, we find that

$$\begin{aligned} \nabla^a F_{ab} &= \nabla^a (E_{ab} - {}^*B_{ab}) \\ &= \nabla^a (E_a v_b - E_b v_a) - \varepsilon_{abcd} \nabla^a B^c v^d \\ &= \nabla^a E_a v_b - \dot{E}_b - \varepsilon_{abcd} \nabla^a B^c v^d \\ &= -4\pi (\rho_0 v_b + K_b). \end{aligned}$$

Thus,

$$\nabla^a E_a = -4\pi \rho_0,$$
$$\dot{E}_b + \varepsilon_{abcd} \nabla^a B^c v^d = 4\pi K_b.$$

In three-vector notation these two equations become

$$\nabla \cdot \boldsymbol{E} = 4\pi \rho_0 \quad \text{and} \quad \dot{\boldsymbol{E}} - \nabla \times \boldsymbol{B} = 4\pi \boldsymbol{K}.$$

Similarly, $\nabla^{a*} F_{ab} = 0$ gives

$$\begin{aligned} \nabla^{a*} F_{ab} &= \nabla^a ({}^*E_{ab} + B_{ab}) \\ &= \varepsilon_{abcd} \nabla^a E^c v^d + \nabla^a B_a v_b - \dot{B}_b = 0. \end{aligned}$$

Thus

$$\varepsilon_{abcd} \nabla^a E^c v^d - \dot{B}_b = 0,$$
$$\nabla^a B_a = 0.$$

In three-vector notation these two equations become

$$\nabla \times \boldsymbol{E} + \dot{\boldsymbol{B}} = 0 \quad \text{and} \quad \nabla \cdot \boldsymbol{B} = 0.$$

8.4 For a perfect fluid,

$$T_{ab} = (\rho + p) u_a u_b - p g_{ab}.$$

The conservation equation, $\nabla^a T_{ab} = 0$, thus gives

$$\nabla^a T_{ab} = (\rho + p) u_a \nabla^a u_b + \nabla^a [(\rho + p) u_a] u_b - \nabla_b p = 0. \tag{S.13}$$

Contracting with u^b, we find that

$$\nabla^a [(\rho + p) u_a] - u^c \nabla_c p = 0.$$

Substituting this back into (S.13), we get

$$(\rho + p)u_a\nabla^a u_b + u^c\nabla_c p u_b - \nabla_b p = 0$$

or, equivalently,

$$(\rho + p)u^a\nabla_a u^b = (g^{bc} - u^b u^c)\nabla_c p.$$

Chapter 9

9.1

$$\tfrac{1}{12}n^2(n^2 - 1).$$

9.3 Use

$$R_{abcd} = -K(g_{ac}g_{bd} - g_{ad}g_{bc}). \tag{S.14}$$

9.4 Since $R_{ab} = 0$,

$$\nabla_a\nabla_b v_c - \nabla_b\nabla_a v_c = -C_{abce}v^e, \tag{S.15}$$

$$\nabla_a\nabla_b v_{cd} - \nabla_b\nabla_a v_{cd} = -C_{abce}v^e{}_d - C_{abde}v_c{}^e. \tag{S.16}$$

Writing $v_a = \nabla_a f$ (which implies $\nabla_{[a}v_{b]} = 0$) and using (S.15), we have

$$\nabla^c\nabla_a\nabla_b v_c - \nabla^c\nabla_b\nabla_a v_c = -\nabla^c C_{abce}v^e,$$

and hence $\nabla^c C_{abce} = 0$ if $\nabla^c\nabla_a\nabla_b v_c = \nabla^c\nabla_b\nabla_a v_c$ or, equivalently,

$$\nabla_c\nabla_a\nabla_b v^c = \nabla_c\nabla_b\nabla_a v^c. \tag{S.17}$$

Since $C^c{}_{bcd} = 0$, equation (S.15) gives $\nabla^c\nabla_b v_c = \nabla_b\nabla^c v_c$ or, equivalently,

$$\nabla_c\nabla_b v^c = \nabla_b\nabla_c v^c. \tag{S.18}$$

We now use (S.18), (S.16), and $C^c{}_{bcd} = 0$ to prove (S.17) and hence $\nabla^c C_{abce} = 0$:

$$\begin{aligned}
&\nabla_a\nabla_c\nabla_b v^c = \nabla_a\nabla_b\nabla_c v^c\\
\Rightarrow\quad &\nabla_c\nabla_a\nabla_b v^c - C_{acbe}\nabla^e v^c - C_{ac}{}^c{}_e\nabla_b v^e = \nabla_b\nabla_a\nabla_c v^c\\
\Rightarrow\quad &\nabla_c\nabla_a\nabla_b v^c - C_{acbe}\nabla^e v^c = \nabla_b(\nabla_c\nabla_a v^c - C_{ac}{}^c{}_e v^e)\\
\Rightarrow\quad &\nabla_c\nabla_a\nabla_b v^c - C_{acbe}\nabla^e v^c = \nabla_b\nabla_c\nabla_a v^c\\
\Rightarrow\quad &\nabla_c\nabla_a\nabla_b v^c - C_{acbe}\nabla^e v^c = \nabla_c\nabla_b\nabla_a v^c - C_{bca}{}^e\nabla_e c^c\\
&\qquad - C_{bc}{}^c{}_e\nabla_a v^e\\
\Rightarrow\quad &\nabla_c\nabla_a\nabla_b v^c - C_{acbe}\nabla^e v^c = \nabla_c\nabla_b\nabla_a v^c - C_{bca}{}^e\nabla_e c^c\\
\Rightarrow\quad &\nabla_c\nabla_a\nabla_b v^c = \nabla_c\nabla_b\nabla_a v^c.
\end{aligned}$$

Thus, $\nabla^c C_{abce} = 0$.

9.5 Any two connections are related according to

$$\hat\nabla_a f = \nabla_a f, \tag{S.19}$$

$$\hat\nabla_a t_b = \nabla_a t_b - \Gamma^e{}_{ab}t_e, \tag{S.20}$$

$$\hat\nabla_a t_{bc} = \nabla_a t_{bc} - \Gamma^e{}_{ab}t_{ec} - \Gamma^e{}_{ac}t_{be}, \tag{S.21}$$

and so on. Since $\hat\nabla_a\hat g_{bc} = \nabla_a g_{bc} = 0$ where $\hat g_{ab} = \Omega^2 g_{ab}$, equation (S.21) (with $\hat g_{ab} = t_{ab}$) gives

$$2\frac{\nabla_c\Omega}{\Omega}g_{ab} - \Gamma_{acb} - \Gamma_{bca} = 0.$$

Permuting indices gives

$$2\frac{\nabla_c\Omega}{\Omega}g_{ab} - \Gamma_{acb} - \Gamma_{bca} = 0,$$

$$2\frac{\nabla_a\Omega}{\Omega}g_{cb} - \Gamma_{cab} - \Gamma_{bca} = 0,$$

$$2\frac{\nabla_b\Omega}{\Omega}g_{ac} - \Gamma_{acb} - \Gamma_{cba} = 0,$$

and hence

$$\Gamma^b{}_{ac} = \gamma_c\delta_a^b + \gamma_a\delta_c^b - \gamma^b g_{ac},$$

where $\gamma_a = \nabla_a\Omega/\Omega$. Substituting this into equations (S.20) and (S.21) now gives

$$\hat{\nabla}_a t_b = \nabla_a t_b - \gamma_a t_b - \gamma_b t_a + g_{ab}\gamma^e t_e, \tag{S.22}$$

$$\hat{\nabla}_a t_{bc} = \nabla_a t_{bc} - 2\gamma_a t_{bc} - \gamma_b t_{ac} - \gamma_c t_{ba} + g_{ab}\gamma^e t_{ec} + g_{ac}\gamma^e t_{be}. \tag{S.23}$$

9.6 (a) Equation (S.22) together with $t^a t_a = 0$ gives

$$\begin{aligned} t^a\hat{\nabla}_a t_b &= t^a(\nabla_a t_b - \gamma_a t_b - \gamma_b t_a + g_{ab}\gamma^e t_e) \\ &= t^a\nabla_a t_b + t^a\gamma_a t_b - t_b\gamma^e t_e \\ &= t^a\nabla_a t_b. \end{aligned}$$

(b) Equation (S.23) together with $t_{ab} = -t_{ba}$ gives

$$\begin{aligned} \hat{\nabla}^b t_{bc} &= \nabla^b t_{bc} - 2\gamma^b t_{bc} - \gamma_b t^b{}_c - \gamma_c t_b{}^b + g_b^b\gamma^e t_{ec} + g_c^b\gamma^e t_{be} \\ &= \nabla^b t_{bc} - 3\gamma^b t_{bc} + 3\gamma^e t_{ec} \\ &= \nabla^b t_{bc}. \end{aligned}$$

Chapter 10

10.1 Since $\hat{g}_{ab} = \Omega^2 g_{ab}$ and $\hat{n}^a = \Omega^{-2}\hat{n}^a$, we have $\hat{n}_a = \hat{g}_{ab}\hat{n}^b = g_{ab}n^b = n_a$. That $\hat{n}_a$ satisfies the geodesic equation $\hat{n}^a\hat{\nabla}_a\hat{n}_b = 0$ now follows directly from Exercise 9.6(a) together with the fact that $n^a\nabla_a n_b = 0$.

10.2 Since $\hat{F}_{ab} = F_{ab}$ and $\nabla^a F_{ab} = 0$, the conclusion $\hat{\nabla}^a\hat{F}_{ab} = 0$ follows directly from Exercise 9.6(b). Similarly, in order to prove $\hat{\nabla}^a {}^*\hat{F}_{ab} = 0$ it is sufficient to show that ${}^*\hat{F}_{ab} = {}^*F_{ab}$. If e_α^a is a right-handed ON basis with respect to g_{ab}, then $\hat{e}_\alpha^a = \Omega^{-1}e_\alpha^a$ is a right-handed ON basis with respect to $\hat{g}_{ab}$. Thus $\hat{\varepsilon}_{abcd}$ is defined by

$$1 = \hat{\varepsilon}_{abcd}\hat{e}_0^a\hat{e}_1^b\hat{e}_2^c\hat{e}_3^d = \Omega^{-4}\hat{\varepsilon}_{abcd}e_0^a e_1^b e_2^c e_3^d,$$

and hence

$$\hat{\varepsilon}_{abcd} = \Omega^4\varepsilon_{abcd}.$$

Furthermore,

$$\hat{F}^{ab} = \hat{g}^{ac}\hat{g}^{bd}F_{cd} = \Omega^{-4}g^{ac}g^{bd}F_{cd} = \Omega^{-4}F^{ab}.$$

Thus,

$${}^*\hat{F}_{ab} = \tfrac{1}{2}\hat{\varepsilon}_{abcd}\hat{F}^{cd} = {}^*F_{ab}.$$

10.3 $\nabla^a C_{abcd} = 0$ follows directly from Exercise 9.4. $\nabla^{a*}C_{abcd} = 0$ follows directly from the second Bianchi identity.

10.4 Equation (10.11) gives

$$R_{ab}v^a v^b = -4\pi(3p + \rho),$$

and hence the Raychaudhuri equation reduces to

$$\dot{\theta} = \tfrac{1}{3}\theta^2 + 4\pi(3p + \rho).$$

If $V = a^3$, then $\dot{V} = -\theta V$ gives $\theta = -3\dot{a}/a$ and hence

$$-3\frac{\ddot{a}}{a} = 4\pi(3p + \rho).$$

10.5 Since $R_{ab}v^a v^b \leq 0$, the Raychaudhuri equation gives

$$\dot{\theta} \geq \frac{\theta^2}{3},$$

and hence

$$(\theta^{-1}) \leq -\tfrac{1}{3}.$$

Thus,

$$\theta^{-1} \leq \theta_0^{-1} - \tfrac{1}{3}t.$$

Since θ_0 is positive, θ must therefore tend to infinity within proper time $3/|\theta_0|$.

Chapter 11

11.1 Using $\hat{m}^a = \Omega^{-1}m^a$, $\hat{l}_a = l_a$, and equation (EE9.1), we have

$$\begin{aligned}\hat{\rho} &= \hat{\bar{m}}^a \hat{m}^b \hat{\nabla}_a \hat{l}_b \\ &= \Omega^{-2}\bar{m}^a m^b \hat{\nabla}_a l_b \\ &= \Omega^{-2}\bar{m}^a m^b(\nabla_a l_b - \gamma_a l_b - \gamma_b l_a + g_{ab}\gamma^e l_e) \\ &= \Omega^{-2}\rho - \Omega^{-3}\nabla^e l_e \\ &= \Omega^{-2}\rho - \Omega^{-3}D\Omega.\end{aligned}$$

11.2 By spherical symmetry we may choose an affine parameter r such that the r = const surfaces are (metric) two-spheres. Furthermore, again by spherical symmetry, there can exist no geometrically defined vector fields tangent to these surfaces – such a vector field would break spherical symmetry. Let $m^a\nabla_a r = 0$, that is, m^a is tangent to r = const. This complex vector does not break spherical symmetry, since it is only defined up to

$$m^a \to e^{i\lambda}m^a.$$

However, if $\sigma \neq 0$, a function α exists such that

$$\sigma\bar{\alpha} = \alpha \quad \text{and} \quad |\alpha| = 1.$$

Since σ is defined up to

$$\sigma \to e^{2i\lambda}\sigma,$$

α is defined up to

$$\alpha \to e^{i\lambda}\alpha$$

and defines a geometric tangent vector to the r = const surfaces by

$$\alpha^a = \bar{\alpha}m^a + \alpha\bar{m}^a.$$

Since this breaks spherical symmetry, $\sigma = 0$.

In the vacuum region, $R_{ab} = 0$, and hence

$$D\rho = \rho^2.$$

Furthermore,

$$DA = -2\rho A.$$

Since $Dr = 1$, these equations imply that r can be chosen such that $A = 4\pi r^2$.

11.4 We have shown that $\hat{l}^a = \Omega^{-2}$, $\hat{m}^a = \Omega^{-1}m^a$, $\hat{\rho} = \rho - \Omega^{-3}D\Omega$, $\hat{\sigma} = \Omega^{-2}\sigma$, and $\hat{D} = \Omega^{-2}D$. Moreover, Exercise 9.6 gives

$$\begin{aligned}\hat{R}_{ab} &= R_{ab} + 2\Omega^{-1}\nabla_a\nabla_b\Omega - 4\Omega^{-2}\nabla_a\Omega\nabla_b\Omega \\ &\quad + g_{ab}(\Omega^{-2}\nabla_c\Omega\nabla^c\Omega + \Omega^{-1}\nabla_c\nabla^c\Omega)\end{aligned}$$

and $\hat{C}_{abcd} = \Omega^2 C_{abcd}$. Thus

$$\begin{aligned}\hat{R}_{ab}\hat{l}^a\hat{l}^b &= \Omega^{-4}(R_{ab}l^a l^b + 2\Omega^{-1}D^2\Omega - 4\Omega^{-2}D\Omega D\Omega), \\ \hat{C}_{abcd}\hat{l}^a\hat{m}^b\hat{l}^c\hat{m}^d &= \Omega^{-4}C_{abcd}l^a m^b l^c m^d.\end{aligned}$$

Using these relations, it is now a simple matter to show that

$$\begin{aligned}D\rho &= \rho^2 + \sigma\bar{\sigma} - \tfrac{1}{2}R_{ab}l^a l^b, \\ D\sigma &= (\rho + \bar{\rho})\sigma + C_{abcd}l^a m^b l^c m^d\end{aligned}$$

imply

$$\begin{aligned}\hat{D}\hat{\rho} &= \hat{\rho}^2 + \hat{\sigma}\bar{\hat{\sigma}} - \tfrac{1}{2}\hat{R}_{ab}\hat{l}^a\hat{l}^b, \\ \hat{D}\hat{\sigma} &= (\hat{\rho} + \bar{\hat{\rho}})\hat{\sigma} + \hat{C}_{abcd}\hat{l}^a\hat{m}^b\hat{l}^c\hat{m}^d.\end{aligned}$$

11.6 Let l^a be tangent to the shear-free congruence and l^a_N tangent to the generators of N. Since l^a is shear-free, equation (11.11) gives $C_{abcd}l^a m^b l^c m^d = 0$. There will exist a generator γ of N on which $l^a = l^a_N$, and hence $C_{abcd}l^a_N m^b_N l^c_N m^d_N = 0$ on γ. Equation (11.10) thus gives

$$D\sigma_N = 2\rho_N\sigma_N$$

on γ. But $\sigma_N = 0$ at the vertex of N, and hence $\sigma_N = 0$ on γ.

Chapter 12

12.1 Writing equation (E9.3) in the form

$$\begin{aligned}\hat{R}''_{ab} &= R'_{ab} + 2\omega^{-1}\nabla'_a\nabla'_b\omega - 4\omega^{-2}\nabla'_a\omega\nabla'_b\omega \\ &\quad + g'_{ab}(\omega^{-2}\nabla'_c\omega\nabla'^c\omega + \omega^{-1}\nabla'_c\nabla'^c\omega),\end{aligned}$$

equation (12.5) follows on letting $g''_{ab} = g_{ab}$, $g'_{ab} = \hat{g}_{ab}$ and $\omega = \Omega^{-1}$.

12.2 For flat spacetime, equation (12.5) gives

$$\hat{R} = 6\Omega^{-1}\hat{\nabla}_a\hat{\nabla}^a\Omega - 12\Omega^{-2}\hat{\nabla}_a\Omega\hat{\nabla}^a\Omega.$$

Since $\nabla_a r\nabla^a r = -1$, we get $\Omega^{-2}\hat{\nabla}_a\Omega\hat{\nabla}^a\Omega = -1$. Using the result of Exercise 9.5, we have

$$\hat{\nabla}_a\hat{\nabla}_b\Omega = \nabla_a\nabla_b\Omega - 2\Omega^{-1}\nabla_a\Omega\nabla_b\Omega + g_{ab}\Omega^{-1}\nabla_e\Omega\nabla^e\Omega,$$

and hence, since $\nabla_a\nabla^a\Omega = 0$,

$$\Omega^{-1}\hat{\nabla}_a\hat{\nabla}^a\Omega = \Omega^{-3}(\nabla_a\nabla^a\Omega + 2\Omega^{-1}\nabla_a\Omega\nabla^a\Omega) = -2.$$

Thus $\hat{R} = 0$.

12.5 From equation (12.12) we have

$$\xi^a\Box\xi_a = R_{ab}\xi^a\xi^b, \tag{S.24}$$

where the **wave operator** $\Box$ is defined by $\Box = \nabla^b\nabla_b$. Thus, using this equation together with (12.16) and the Killing equation, we get

$$\begin{aligned}\Box V^2 = \nabla^b\nabla_b V^2 &= 2\nabla^b(V\nabla_b V) \\ &= -2\nabla^b(\nabla_b\xi_a\xi^a) \\ &= -2(\nabla^b\nabla_b\xi_a\xi^a + \nabla_b\xi_a\nabla^b\xi^a)\end{aligned}$$

$$= -2(\Box\xi_a\xi^a)$$
$$= -2R_{ab}\xi^a\xi^b. \tag{S.25}$$

Note that in matter-free regions where T_{ab} and hence R_{ab} vanish, this gives

$$\Box V^2 = 0. \tag{S.26}$$

Chapter 13

13.1 Equations (13.5) and (13.19) give

$$dt = du - du',$$
$$2dr = -\left(1 - \frac{2m}{r}\right)(du + du').$$

Starting from the form of the metric given by equation (13.21) and using these relations, we get

$$g = \left(1 - \frac{2m}{r}\right) du'^2 + 2\,du'\,dr - r^2(d\theta^2 + \sin^2\theta\; d\varphi^2), \tag{S.27}$$

$$g = -\left(1 - \frac{2m}{r}\right) du\,du' - r^2(d\theta^2 + \sin^2\theta\; d\varphi^2), \tag{S.28}$$

$$g = \left(1 - \frac{2m}{r}\right) dt^2 - \left(1 - \frac{2m}{r}\right)^{-1} dr^2 - r^2(d\theta^2 + \sin^2\theta\; d\varphi^2). \tag{S.29}$$

13.2 Using equation (13.5), we have

$$D'u = l'^a\nabla_a u = l'^a l_a = -2 - \frac{4m}{r - 2m}.$$

Writing $u = u(r)$ and using $D'r = 1$, this integrates to give

$$u = -2r - 4m\ln|r - 2m| + C.$$

13.4 Let v^a be the four-velocity of the orbit. Since v^a is future-pointing and r^a is past-pointing in II, $v^a r_a < 0$ and hence the radius of the orbit must decrease with time. No closed orbit can therefore exist in II. Similarly for II$'$.

13.5 For a radial orbit, $w^a = a\xi^a + br^a$. Since $\xi_a\xi^a = V^2$ and $r_a r^a = -V^2$, where $V^2 = 1 - 2m/r$, the fact that $w_a w^a = 1$ implies

$$V^2(a^2 - b^2) = 1.$$

Since $\xi^a r_a = 0$, the energy of the orbit (which is constant) is given by $E = w^a\xi_a = V^2 a$. Moreover $\dot{r} = w^a r_a = -V^2 b$. From these relations a simple calculation gives

$$\dot{r}^2 = E^2 - 1 + \frac{2m}{r}.$$

13.6 Let $w^a = a\xi^a + br^a$ be O's four-velocity. Since he starts from rest at infinity and travels radially *inwards*, $E = w^a\xi_a = V^2 a = 1$ and $w^a r_a = -V^2 b < 0$ (i.e. $b > 0$). For $r = r_0$ the results of the previous exercise give $a = V_0^{-2}$ and $b^2 = 2mr_0^{-1}V_0^{-4}$. An incoming photon with frequency ω_0 at infinity has four-momentum $p_a = -\omega_0 l'_a$. The received frequency is thus

$$\omega = p_a w^a = -\omega_0(-a + b).$$

(Here we have used $l'^a r_a = 1$ and $l'^a\xi_a = -1$.) Similarly, an outgoing photon with frequency ω at infinity has four-momentum $p_a = \omega l_a$. The transmitted frequency is thus

$$\omega_0 = p_a w^a = \omega(a+b),$$

and hence

$$\omega = \omega_0(a+b)^{-1}.$$

(Here we have used $l^a r_a = 1$ and $l^a \xi_a = 1$.) Note that $\omega \to 0$ as $r \to 2m$.

13.7 On the surface of the star, $v^a = \xi^a / V$ is the four-velocity of a stationary observer and $e^a = r^a / V$ is a unit vector pointing vertically upwards. If w^a is the four-velocity of a particle projected vertically upwards, then its speed v_0 relative to the surface is given by

$$w^a = \gamma(v^a + v_0 e^a),$$

where $\gamma = (1 - v_0^2)^{-1/2}$. If the particle just escapes to infinity, then

$$E = \xi_a w^a = \gamma V = 1$$

and hence

$$V^2 = 1 - \frac{2m}{r_0} = 1 - v_0^2$$

which gives $v_0^2 = 2m/r_0$.

Chapter 14

14.1 Since l'^a and r^a are well defined and linearly independent on H, we may write the four-velocity of the observer as

$$v^a = al'^a + br^a.$$

Since he starts from rest at infinity,

$$E = \xi_a v^a = -a = 1,$$

where we have used $\xi_a l'^a = -1$ and $\xi_a r^a = 0$. Since $l'^a r_a = 1$ and $r^a r_a = 0$ on H, we have

$$1 = v^a v_a = 2ab$$

and hence $b = \frac{1}{2}$. The four-velocity of the photon is $p_a = -\omega_0 l'_a$, because $p_a \xi^a = \omega_0$ and ξ^a tends to the four-velocity of a stationary observer at infinity. The received frequency is thus $\omega = p_a v^a = w_0/2$.

14.2 Since $\xi^a \nabla_a u = 1$ and ξ^a tends to the four-velocity of a stationary observer at infinity, u is the proper time for such an observer. But $u \to \infty$ as H is approached, and hence the particle will never appear to reach $r = 2m$, according to a stationary observer at infinity.

14.3 Complete l^a to form a null tetrad $(n^a, l^a, m^a, \bar{m}^a)$. Since $l_a l^a = 0$ and $l^a \nabla_a l_b = 0$,

$$(l^a n^b + n^a l^b)\nabla_a l_b = 0.$$

Thus

$$\begin{aligned}\nabla^a l_a &= (l^a n^b + n^a l^b - m^a \bar{m}^b - \bar{m}^a m \hat{b} \nabla_a l_b) \\ &= -\frac{1}{r + ia\cos\theta} - \frac{1}{r - ia\cos\theta} \\ &= -\frac{2r}{\Sigma}.\end{aligned}$$

14.5 Equation (14.6) shows that the red-shift factor is

$$V^2 = \xi_a \xi^a = 1 - \frac{2mr}{r^2 + a^2\cos^2\theta}.$$

If ω_0 is the frequency of light emitted from the surface of the star, the received frequency is given by

$$\omega = \omega_0 V.$$

The north and south poles will thus appear redder than the equator.

14.6 Let $n^a = Vl^a$, $n'^a = -Vl'^a$, and $v^a = V^{-1}\xi^a$, where $V^2 = \xi_a\xi^a$. Then v^a is the four-velocity of a stationary observer, and $n^a v_a = n'^a v_a = 1$ and $n^a n'_a = 2$. Thus,

$$n^a = v^a + e^a \quad \text{and} \quad n'^a = v^a + e'^a,$$

where e^a and e'^a are unit vectors orthogonal to v^a. Therefore,

$$\begin{aligned} n^a n'_a &= 1 + e^a e'_a = 1 - \mathbf{e}\cdot\mathbf{e}' = 1 - \cos\phi \\ &= -V^2 l^a l'_a = 2V^2\Sigma\Delta^{-1} \\ &= 2\frac{r^2 - 2mr + a^2\cos^2\theta}{r^2 - 2mr + a^2}. \end{aligned}$$

14.7 As a (round) disk! This follows from Exercise 11.6.

14.8 For Schwarzschild

$$K_{ab} = r^2\left[\left(1 - \frac{2m}{r}\right)l_{(a}l'_{b)} + g_{ab}\right].$$

That K_{ab} is a Killing tensor follows from the equations

$$\nabla_a l_a = \frac{m}{r^2} l_a l_b + \left(1 - \frac{2m}{r}\right)l_{(a}l'_{b)} + \frac{1}{r}g_{ab},$$

$$\nabla_a l'_a = \frac{m}{r^2} l'_a l'_b + \left(1 - \frac{2m}{r}\right)l_{(a}l'_{b)} + \frac{1}{r}g_{ab},$$

$$2r_a = \left(1 - \frac{2m}{r}\right)(l_a + l'_a)$$

(see Exercise 13.8).

In terms of a null tetrad with $2n^a = -(1 - 2m/r)l'^a$ and m^a tangent to the $r = $ const two-spheres, we have

$$K_{ab} = r^2(-l_a n_b - n_a l_b + g_{ab}),$$

and since

$$g_{ab} = l_a n_b + n_a l_b - m_a\bar{m}_b - \bar{m}_a m_b,$$

we have

$$K_{ab} = -r^2(m_a\bar{m}_b + \bar{m}_a m_b) = r^2 k_{ab},$$

where k_{ab} is the intrinsic metric of the $r = $ const two-spheres.

14.9 Writing

$$w^a = a\xi^a + br^a + c\chi^a,$$

we have

$$1 = w_a w^a = \left(1 - \frac{2m}{r}\right)(a^2 - b^2) - c^2 r^2,$$

$$J = w^a\chi_a = cr^2,$$

$$K = K_{ab}w^a w^b = r^2\left[\left(1 - \frac{2m}{r}\right)(-a^2 + b^2) + 1\right],$$

and hence $K = -J^2$.

Chapter 16

16.1 For a static universe with k equal to 1, 0, or -1, the curvature of the $\hat{t} = \text{const}$ cross sections is given by $K = k$. Thus,

$$\hat{g} = d\hat{t}^2 - \frac{dr^2}{1 - kr^2} + r^2(d\theta^2 + \sin^2\theta\, d\varphi^2).$$

But $g = a^2 g$ [equation (16.12)] and $a\, d\hat{t} = dt$ [equation (16.14)], and hence

$$g = dt^2 - \frac{a^2\, dr^2}{1 - kr^2} + r^2(d\theta^2 + \sin^2\theta\, d\varphi^2).$$

This is *not* a global result. If it were, then the $t = \text{const}$ crosssections would be restricted to have topology S^3 (for $k = 1$) or $\mathbb{R}^3$ (for $k = 0$ or -1), but a space of constant curvature can have many other topologies. For example, P^3, defined by identifying antipodal points of S^3, is a space of constant curvature. For an interesting discussion of cosmic topology, see Lachieze-Ray and Luminet (1995).

16.3 Starting with a conformally related static universe with constant curvature $k = 1$, equation (16.11) gives $\hat{R}_{ab} = 2\hat{h}_{ab}$ where $\hat{h}_{ab} = \hat{g}_{ab} - \hat{v}_a\hat{v}_b$. Let $\hat{l}^a$ be tangent to the generators of the null cone N and such that

$$\hat{l}^a \nabla_a \hat{t} = \hat{l}^a \hat{v}_a = 1.$$

The Raychaudhuri equation (11.9) gives $\hat{D}\hat{\rho} = \hat{\rho}^2 + 1$ with solution

$$\hat{\rho} = -\cot(\hat{t} - \hat{t}_0),$$

where the constant of integration has been chosen such that $\hat{\rho}$ is infinite at the vertex of N. If $\hat{A}$ is the area of the cross sections of N, then $\hat{D}\hat{A} = -2\hat{A}$ with solution

$$\hat{A} = 4\pi \sin^2(\hat{t} - \hat{t}_0).$$

Since $g_{ab} = a^2 \hat{g}_{ab}$, the required area is given by

$$A = 4\pi a^2 \sin^2(\hat{t} - \hat{t}_0).$$

Note that N converges to a new vertex in a $\hat{t}$-interval of π.

16.6 This follows directly from equation (16.20).

16.7 Since $d\hat{t} = a\, dt$, equation (16.20) implies that the life span of the universe in $\hat{t}$-seconds is given by

$$\Delta\hat{t} = \pi C/2.$$

If the universe has topology S^3, then a null cone must have time to reconverge *twice* if the observer is to see the back of his head: once at the antipodal point and once again on the observer's world line. In Exercise 16.3 we showed that a null cone reconverges in a $\hat{t}$-interval equal to π. Thus, if the observer sees the back of his head, $2\pi < \Delta\hat{t}$, and hence $4\pi < C$. If, on the other hand, the topology is P^3, then the null cone need only reconverge once, and hence $2\pi < C$.

Bibliography

J. D. Barrow and F. I. Tipler. 1986. *The anthropic cosmological principle.* Clarendon Press, Oxford.

J. D. Bekenstein. 1974. The generalized second law of thermodynamics in black hole physics. *Phys. Rev. D* 9:3292.

M. Berry. 1976. *Principles of cosmology and gravitation.* Cambridge University Press.

H. Bondi. 1961. *Cosmology.* Cambridge University Press.

H. Bondi. 1980. *Relativity and common sense.* Dover, New York.

C. H. Brans. 1994. Exotic smoothness and physics. *J. Math. Phys.*, 35:5494–5506.

H. A. Buchdahl. 1975. *Twenty lectures on thermodynamics.* Pergamon Press, Australia.

S. Chandrasekhar. 1957. *Stellar structure.* Dover, New York.

S. Chandrasekhar. 1983. *The mathematical theory of black holes.* Clarendon Press, Oxford.

C. Choquet-Bruhat, Y. De Witt-Morette, and M Dillard Bleick. 1977. *Analysis, manifolds and physics.* North-Holland, Amsterdam.

M. Crampin and F. A. E. Pirani. 1986. *Applicable differential geometry.* Cambridge University Press.

P. C. W. Davies. 1974. *The physics of time asymmetry.* Berkley.

Ray d'Inverno. 1992. *Introducing Einstein's relativity.* Clarendon Press, Oxford.

W. G. Dixon. 1978. *Special relativity, the foundation of modern physics.* Cambridge University Press.

R. Engelking. 1968. *Outline of general topology.* North-Holland and PWN, Amsterdam.

M. Gardner. 1967. *The ambidextrous universe.* Allen Lane, The Penguin Press, London.

P. R. Halmos. 1974. *Finite dimensional vector spaces.* Springer-Verlag, New York.

S. W. Hawking. 1975. Particle creation by black holes. *Comm. Math. Phys.*, 43:199.

S. W. Hawking and G. F. R. Ellis. 1973. *The large scale structure of space-time.* Cambridge University Press.

S. Hawking and R. Penrose. 1996. *The nature of space and time.* Princeton University Press, Princeton.

N. S. Hetherington. 1993. *Encyclopedia of cosmology.* Garland Publishing, New York.

L. P. Hughston and K. P. Tod. 1990. *An introduction to general relativity.* Cambridge University Press.

J. N. Islam. 1992. *Introduction to mathematical cosmology*. Cambridge University Press.

A. Komar. 1959. Covariant conservation laws in general relativity. *Phys. Rev.*, 113:934–936.

M. Lachieze-Ray and J. P. Luminet. 1995. Cosmic topology. *Phys. Reports*, 254: 135–214.

D. Lovelock. 1972. The four-dimensionality of space and the Einstein tensor. *J. Math. Phys.*, 13:874–876.

J. V. Narlikar. 1983. *Introduction to cosmology*. Jones and Bartlett, Portola Valley, California.

R. Penrose. 1959. The apparent shape of a relativistically moving sphere. *Proc. Camb. Phil. Soc.*, 55:137–139.

R. Penrose. 1968. Structure of space-time. In *Battelle Rencontres, 1967 Lectures in Mathematics and Physics*, eds. C. M. DeWitt and J. A. Wheeler. Benjamin, New York.

R. Penrose. 1969. Gravitational collapse: The role of general relativity. *Rev. Nuovo Cimento*, 1:252–276.

R. Penrose. 1990. *The emperor's new mind*. Oxford University Press.

R. Penrose and W. Rindler. 1986. *Spinors and space-time. Vols. 1 and 2.* Cambridge University Press.

R. D. Reed and R. R. Roy. 1971. *Statistical physics for students of science and engineering*. Intext Educational Publishers, San Francisco.

R. Schoen and S. T. Yau. 1979. The positivity of the mass of a general space-time. *Phys. Rev. Lett.*, 44:1457–1459.

R. Schoen and S. T. Yau. 1983. The existence of a black hole due to condensation of matter. *Comm. Math. Phys.*, 90:575.

N. E. Steenrod. 1951. *The topology of fibre bundles*. Princeton University Press, Princeton.

R. M. Wald. 1984. *General relativity*. University of Chicago.

R. M. Walker and R. Penrose. 1970. On quadratic first integrals of the geodesic equations for type [22] spacetimes. *Comm. Math. Phys.*, 18:265–274.

S. Weinberg. 1972. *Gravitation and cosmology*. Wiley, New York.

S. Weinberg. 1993. *The first three minutes*. Flamingo, London.

E. Witten. 1981. A new proof of the positive energy theorem. *Comm. Math. Phys.*, 80:381–402.

Index

Printed in the United States
By Bookmasters